Reasonable Books = Reasonable Prices

Irina Kleyman

Exceptional for Home Schooling

Math Help Grade 7

First Edition

Reasonable Books = Reasonable Prices

Irina Kleyman

Exceptional for Home Schooling

Math Help Grade 7

I dedicate this book to my children and grandchildren.

Acknowledgments
Special thanks to Michael G LaFratta, Vladimir Kleyman, who has assisted in bringing all 10 books to market.

Author's note: In the many years that I've taught mathematics, I have been informed by many of my students that my methods for teaching have helped them tremendously in learning math. I have written these books in order to continue to help others learn math effectively.

Math Help Grade 1
Math Help Grade 2
Math Help Grade 3
Math Help Grade 4
Math Help Grade 5
Math Help Grade 6
Math Help Grade 7
Math Help Grade 8
Math Help for College (Summary Grades 1 – 5)
Math Help for University (Basic, Algebra, Geometry)

ISBN: 978-1-7349699-6-2 Paperback

Syllabus Math 5, 6, 7, 8
Grade 5#1–5, Grade 6 # 6–10, Grade 7 # 11–15, Grade 8 # 16–20

#	Grades	Topics: Advanced Problems (A), Pages (p)
	Chapters	Basic Problems (B), Chapter (C), Lesson (L)
1	5-1	Four–digit numbers (+, -) (B)
2	5-2	Multiplying numbers (B)
3	5-3	Dividing numbers (B)
4	5-4	Four operations with numbers (B)
5	5-5	Finding unknown (A-5)
	Grade 5	Final Exam 5 B, C 5, L16 #1-10 all, p.150
6	6-1	Many-digit numbers (B)
7	6-2	Multiplying numbers (B)
8	6-3	Dividing numbers (B)
9	6-4	Four operations with numbers (B)
10	6-5	Finding unknown (A-6)
	Grade 6	Final Exam 6 B, C 5, L 16 #1-10 all, p.196
11	7-1	Whole numbers (B)
12	7-2	Properties of whole numbers (B)
13	7-3	Decimals (B)
14	7-4	Fractions and mixed numbers (B)
15	7-5	Converting fractions and decimals (B)
	Grade 7	Final Exam 7 B, C 5, L 12 #1-10 all, p.209-210
16	8-1	Rational numbers (B)
17	8-2	Equations (B)
18	8-3	Geometry (B)
19	8-4	Word Problems (B)
20	8-5	Summary of Basic Math (B)
21	8-6	Extra Review (B)
	Grade 8	Final Exam 8 B, C 5, L 14 #1-10 all, p.281-282

Syllabus includes topics, what students need to know.
Middle School can start from 5th, 6th, or 7th grade. Do not worry about school. Choose your grade, work hard and you have good knowledge.

Grade 7 Chapter 1

Whole Numbers

www.mathbooksbyirina.com

Parents, Tutors, Students Middle Schools, High Schools, Colleges, University can help primary school students. Please read slowly math books to primary school students.

Grade 7 Chapter 1 Contents
Whole numbers
Lesson 1 –10
Basic Problems

Work with lesson 1, read rules and examples, 2) Solve the problems from contents, 3) Check answers in the end of chapter 1, 4) Work with the next lessons 2, 3, 4, 5, ..., 5) Solve Test 1 A, 6) Grading Scale: A =10 points, B=9points, C=8 points, D=7 points, F=0-6 points, 7) If you solve 10 problems correctly, you get an A, 8) If have an A, you can move on to next chapter, 9) If your grade is lower than an A, work with the mistakes, then solve Test 1B, 10) When you finish all of the chapters, solve Final Exam in last chapter.

#1 – 20 odd or # 1, 3, 5, 7, 9, 11, 13, 15, 17, 19

Lesson 1 Solve the Pretest 1 # 1, # 2, # 3, # 4, # 5
College Placement Test # 1, # 2, # 3, # 4, # 5, # 6, # 7, # 8, # 9, # 10
Lesson 2 Solve the problems Review 1 A # 1, # 2, # 3, # 4, # 5
Solve the problems Review 1 B # 1, # 2, # 3, # 4, # 5
Lesson 3 Comparison and Rounding of whole numbers.
Solve the problems Review 1 A # 1, 3, 5, 7, 9, 11, 13, 15, 17, 19
Review 1 B # 1, 3, 5, 7, 9, 11, 13, 15, 17, 19
Lesson 4 Addition and Subtraction of whole numbers.
Solve Review 1 A # 4 (a, b, c, d, e), # 8 (a, b, c, d, e)
Solve Review 1 B # 4 (a, b, c, d, e), # 8 (a, b, c, d, e)
Lesson 5 Multiplication of whole numbers.
Memorize Table 1. Solve Review 1 A # 1–# 10, Review 1 B # 1-# 10
Lesson 6 Division of whole numbers. Review 1 A # 1 – 10
Lesson 7 Complete the table of division
Solve Review 2 A # 1 – # 20 odd, Review 2 B # 1 – # 20 odd
Lesson 8 Solve the Review 1 A # 1 – # 10 all
Solve the Review 1 B # 1 – # 10 all
Lesson 9 Solve Test 1 A # 1 – # 10 all
Solve Test 1 B # 1 – # 10 all
Lesson 10 Finding unknown (optional)
Review 1 A # 1 – # 20 odd, Review 1 B # 1 – # 20 odd
Review 2 A # 1 – # 20 odd, Review 2 B # 1 – # 20 odd
Answers for Chapter 1

Grade 7 Chapter 1 Lesson 1
Pretest 1 (25 minutes)

1) **Multiply** $\begin{array}{r} 807 \\ \underline{\times 504} \end{array}$

1)_______

2) **Divide** $245427 \div 403$

2)_______

3) **Evaluate** $420420 \div 70 \times 2$

3)_______

4) **Evaluate** $640640 - 320320 \div 4 \times 5$

4)_______

5) **Evaluate** $180 \div 6 \times 2 - 66 \div 6 \times 3$

5)_______

How do you feel about Math?
a) Love b) Like c) Nothing d) Do not Like e) Hate

Grade 7 Chapter 1 Lesson 1
College Placement Test (50 minutes)

Read directions carefully. **Show all steps for full credits.**

Do only 1 operation per 1 step. **Perform the indicated operations.**

1) $\{5\times5^2 - 2^3[2^2 + (1+2)^2]\}+2^3$

2) $4.9 \times 7 - 1.8 \times 7 + 6.9 \times 7$

3) $\dfrac{4-\frac{1}{4}}{3+\frac{1}{8}}$

4) $-2\frac{6}{7} \div 1\frac{3}{7} - 2\frac{1}{7} \div 1\frac{3}{7}$

5) Simplify $4(\frac{1}{2}c - \frac{1}{4}d) - \frac{2}{3}(6c+3d)$

6) Solve the equation $0.3y + 0.2(2y-3) = 0.8 - 0.7(y-2)$

7) Find the perimeter of right triangle if it hypotenuse is $5\frac{1}{10}$ ft and another leg is $2\frac{2}{5}$ ft.

8) Find the area of trapezoid, if it height is $4\frac{2}{3}$ m, bases are $8\frac{1}{2}$ m and $6\frac{1}{2}$ m.

9) A bus drove 75 mi at a speed of 30 mi/h.
How much time bus spend?

10) Of 32 children, 62.5% were boys. How many were girls?

Grade 7 Chapter 1 Lesson 2
Review 1 A (25 minutes)

Read directions carefully. **Answers must be in the indicated place.**
Show all steps for full credits. **Do only 1 operation per 1 step.**
Solve 1 problem 5 minutes. **Each problem is worth 5 points.**

1) **Multiply** $\begin{array}{r} 706 \\ \underline{\times 605} \end{array}$

1)_______

2) **Divide 243714 ÷ 302**

2)_______

3) **Evaluate** $240240 \div 80 \times 3$

3)_______

4) **Evaluate** $560560 - 480480 \div 8 \times 5$

4)_______

5) **Evaluate** $120 \div 3 \times 2 - 60 \div 4 \times 3$

5)_______

Grade 7 Chapter 1 Lesson 2
Review 1 B

Read directions carefully. Answers must be in the indicated place.
Show all steps for full credits. Do only 1 operation per 1 step.
Solve 1 problem 5 minutes. Each problem is worth 5 points.

1) Multiply

$$\begin{array}{r} 605 \\ \underline{\times 403} \end{array}$$

1)_______

2) Divide $182508 \div 201$

2)____

3) Evaluate $180180 \div 90 \times 2$

3)_______

4) Evaluate $480480 - 360360 \div 6 \times 5$

4)_______

5) Evaluate $150 \div 5 \times 2 - 60 \div 5 \times 3$

5)_______

Grade 7 Chapter 1 Lesson 3
Operations with Whole Numbers
Comparison and Rounding of Whole Numbers

Arithmetic based on mathematics, which studies properties of operations with numbers. All numbers are made up of the digits: 0, 1, 2, 3, 4, 5, 6, 7, 8, and 9. Basic operations are: addition, subtraction, multiplication, and division. Natural (counting) numbers are numbers that use for calculations 1, 2, 3 ... Whole numbers are natural numbers and zero: 0, 1, 2, 3, 4, 5 ... Whole numbers can have represented with points on a number line

Odd numbers are 1, 3, 5, 7, 9, 11, 13, 15...
Even numbers are 0, 2, 4, 6, 8, 10, 12, 14, 16, 18 ...
The first operation with numbers is addition, which combines two or more numbers together.

The first number in addition is called the addend	**2**
The second number in addition is called the addend	**3**
The result of addition is called the sum	**5**
Addend + Addend = Sum	**$2 + 3 = 5$**

The second operation with numbers is subtraction, which means to take something away.

The first number in subtraction is called the minuend	**5**
The second number in subtraction is called the subtrahend	**2**
The result of subtraction is called the difference	**3**
Minuend – Subtrahend = Difference	**$5 - 2 = 3$**

Rounding

Using an approximate value for number called rounding.
A rounded number is always rounded to a given place value
Numbers can round to any digit: one, ten, and hundred ...
If the digit to the right of the given place value is less than 5 (0, 1, 2, 3, 4), then that digit and all digits to the right are replaced by zeros. Rounding a number to the ten: $283 \approx 280$
If the digit to the right of the given place value is greater than or equal to 5 (5, 6, 7, 8, 9), increase the digit in the given place value by 1, and replace all other digits to the right by zeros. Rounding numbers to the ten: $238 \approx 240$

Grade 7 Chapter 1 Lesson 3
Comparison numbers

Determine which of number greater than (>) or smaller than (<) another number, or numbers are equal (=). 2 < 7   8 > 3   5 = 5

Any whole number greater than zero   6 > 0

Number zero smaller than any whole number   0 < 5

To compare numbers, compare digits from left to right

1234 = 1234   1234 < 1235 (4 < 5)   1238 > 1236   8 > 6)

Review 1 A

Round the numbers to the nearest ten (to the 10)

1) 123
2) 234
3) 345
4) 456
5) 567
6) 678
7) 789
8) 898
9) 987
10) 876

Round the numbers to the nearest hundred (to the 100)

11) 1234
12) 2345
13) 3456
14) 4567
15) 5678
16) 6789
17) 7898
18) 8987
19) 9876
20) 8765

Review 1 B

Compare the numbers

(Put Symbols <, >, or = between numbers)

1) 1 ? 14
2) 345 ? 23
3) 456 ? 1234
4) 12345 ? 2345
5) 2345 ? 12345
6) 23456 ? 9876
7) 3456 ? 23456
8) 34567 ? 8765
9) 9876 ? 9876
10) 8765 ? 8765
11) 9876 ? 9080
12) 1234 ? 1325
13) 1001 ? 1001
14) 1234 ? 1543
15) 9876 ? 9767
16) 1234 ? 1243
17) 3456 ? 3465
18) 3465 ? 3456
19) 1243 ? 1234
20) 2345 ? 2354

Grade 7 Chapter 1 Lesson 4
Addition and Subtraction of Whole Numbers

The standard form of numbers is 7, 76, and 765
The expanded form of number is $76 = 70 + 6 = 7\times10 + 6$
$765 = 700 + 60 + 5 = 7\times 100 + 6\times10 + 5$
For large numbers, commas used to separate digits into groups of three called periods. Each period has three digits. The one-digit number is 7, the two-digit number is 76, and the three-digit number is 765.
The four-digit number is 7,654; the five-digit number is 76,543; and six-digit number is 765,432. The first period that has three digits is called ones, for example, in the number 765. The place value of the first period, read from left to right, are: hundreds is 7, tens is 6, ones is 5.
The second period has six digits and called thousands, for example the number 765, 432. The place value of the second period thousands, read from left to right are: hundred-thousand is 7, ten-thousand is 6, thousands is 5, hundreds is 4, tens is 3, and ones is 2.
Six-digit number contains six digits: 100,000; 100,001; ...999,999. The least six-digit number is 100,000; greatest six-digit number is 999,999.The third new period has nine digits and is called millions: 987, 654, 321

The place value of the third period, read from left to right are: hundred million is 9, ten-millions is 8, millions is 7, hundred thousand is 6, ten-thousands is 5, thousands is 4, hundred is 3, tens is 2, and ones is 1.
Seven- digit numbers contain seven digits: 1,000,000 ... 9,999,999. The name of seven- digit numbers from left to right are millions, Hundred thousand, ten-thousand, thousand, hundred, ten, one.
A one-digit number is always less than a two, three, four, five, six or Seven- digit numbers
A seven-digit number is always greater than a six, five, four, three, two, or one-digit numberю. To compare numbers, compare digits from left to right: Hundred thousand, ten-thousand, thousand, hundred, ten, and one: $123456 = 123456$ $123456 < 123459$ $(6 < 9)$ $123458 > 123450$ $(8 > 0)$

Grade 7 Chapter 1 Lesson 4
Properties of Addition and Subtraction

Letter can represent any numbers. A and B can be any numbers.
Changing the order of two addends does not change the sum.

A + B = B + A   2000 + 3000 = 5000   3000 + 2000 = 5000

The sum of any number and zero is the same number.

A + 0 = 0 + A = A   5000 + 0 = 5000   0 + 5000 = 5000

The difference between any number and zero is the same number.

A – 0 = A   5000 – 0 = 5000

The difference between two of the same numbers is zero.

A – A = 0   5000 – 5000 = 0

To check addition use subtraction

200 + 300 = 500   500 – 200 = 300   500 – 300 = 200

To check subtraction use addition or subtraction.

500 – 200 = 300   500 – 300 = 200   200 + 300 = 500

Adding six-digit numbers: (work from right to left)
add the ones, add the tens, add the hundred, add the thousand, add the ten thousand and add the hundred thousand.
Subtracting six- digit numbers: (work from right to left)
subtract the ones, subtract the tens, and subtract the hundreds, subtract the thousands; subtract the ten thousand, and subtract the hundred-thousands

```
                                1 1 1 1      8 17 16 15 14 13
 654321      777777             456789         9 8 7 6 5 3
+123456     -123456           + 345678      -  7 8 9 7 8 9
 777777      654321             802467         1 9 7 8 6 4
```

Addition of seven- digit numbers: (work from right to left)
add the ones, add the tens, add the hundreds, add the thousands, add the ten thousand, add the hundred thousand, add the millions.
Subtraction of seven- digit numbers: (work from right to left)
subtract the ones, subtract the tens, subtract the hundreds, subtract the thousands; subtract the ten thousand, subtract the hundred, subtract thousand, subtract the millions

```
     1  1             10 12
  1020304           6081012
+ 5060708         - 1020304
  6081012           5060708
```

Grade 7 Chapter 1 Lesson 4
Addition and subtraction numbers
Remember

10 = 9 + 1 = 8 + 2 = 7 + 3 = 6 + 4 = 5 + 5
11 = 9 + 2 = 8 + 3 = 7 + 4 = 6 + 5
12 = 9 + 3 = 8 + 4 = 7 + 5 = 6 + 6
13 = 9 + 4 = 8 + 5 = 7 + 6
14 = 9 + 5 = 8 + 6 = 7 + 7

15 = 9 + 6 = 8 + 7
16 = 9 + 7 = 8 + 8
17 = 9 + 8
18 = 9 +9
19 = 10 + 9

Review 1 A
Add the numbers

	a	b	c	d	e
1)	191900 +119100	282800 +128200	373700 +137300	464600 +146400	595900 +159500
2)	292900 +129200	383800 +138300	474700 +147400	565600 +156500	686800 +168600
3)	393900 +139300	484800 +148400	575700 +157500	696900 +169600	797900 +179700
4)	494900 +149400	585800 +158500	676700 +167600	787800 +178700	898900 + 89800

Subtract the numbers

	a	b	c	d	e
5)	311000 -191900	411000 -282800	511000 - 373700	611000 -464600	755400 -595900
6)	422100 -292900	522100 -383800	622100 -474700	722100 -565600	855400 - 686800
7)	533200 -393900	633200 -484800	733200 -575700	866500 -696900	977600 - 797900
8)	644300 -494900	744300 -585800	844300 -676700	966500 -787800	988700 - 898900

Grade 7 Chapter 1 Lesson 4
Review 1 B
Add the numbers

	a	b	c	d	e
1)	1919000 +1191000	2828000 +1282000	3737000 +1373000	4646000 +1464000	5959000 + 1595000
2)	2929000 +1292000	3838000 +1383000	4747000 +1474000	5656000 +1565000	6868000 +1686000
3)	3939000 +1393000	4848000 +1484000	5757000 +1575000	6969000 +1696000	7979000 +1797000
4)	4949000 +1494000	5858000 +1585000	6767000 +1676000	7878000 +1787000	8989000 + 898000

Subtract the numbers

	a	b	c	d	e
5)	3110000 -1919000	4110000 -2828000	5110000 -3737000	6110000 -4646000	7554000 - 5959000
6)	4221000 -2929000	5221000 -3838000	6221000 -4747000	7221000 -5656000	8554000 - 6868000
7)	5332000 -3939000	6332000 -4848000	7332000 -5757000	8665000 -6969000	9776000 - 7979000
8)	6443000 -4949000	7443000 -5858000	8443000 -6767000	9665000 -7878000	9887000 - 8989000

Grade 7 Chapter 1 Lesson 5
Multiplication of Whole Numbers

The third operation with numbers is multiplication × or ·
Multiplication is the operation of repeated addition of the same numbers $2 \times 3 = 2 + 2 + 2 = 6$ **or** $2 \cdot 3 = 2 + 2 + 2 = 6$

The first number in multiplication is called the factor 2
The second number in multiplication is called the factor 3
The result of multiplication is called the product 6
Factor × Factor = Product $2 \times 3 = 6$

Properties of Multiplication

Changing the order of two factors does not change the product
$AB = BA$ $2 \times 3 = 2 + 2 + 2 = 6$ $3 \times 2 = 3 + 3 = 6$

The product of any number and zero is zero
$A \times 0 = 0 \times A = 0$ $5 \times 0 = 0$ $0 \times 5 = 0$

The product of any number and one is the same number
$A \times 1 = 1 \times A = A$ $5 \times 1 = 5$ $1 \times 5 = 5$

Memorize Rules

$A + 0 = A$	$0 + A = A$	$A - 0 = A$	$A - A = 0$
$A \times 1 = A$	$1 \times A = A$	$A \times 0 = 0$	$0 \times A = 0$

Multiplying Numbers by 10, 100 …

To multiply numbers by 10, add zero to the end of a number:
$5 \times 10 = 50$
To multiply numbers by 100, add two zeros to end of a number:
$5 \times 100 = 500$

Table 1
Multiplication and Division

$A + 0 = A$ $0 + A = A$ $A - 0 = A$ $A - A = 0$

$A \times 1 = A$ $1 \times A = A$ $A \div 1 = A$ $A \div A = 1$

$A \times 0 = 0$ $0 \times A = 0$ $0 \div A = 0$ $A \div 0 = \emptyset$

$A + B = B + A$ $A \times B = B \times A$ $\emptyset$ - no solutions

Number 5 from the left column multiply by number 4 from first row to get answer 20. $5 \times 4 = 20$ $20 \div 5 = 4$ $20 \div 4 = 5$

Memorize Table

×	1	2	3	<u>4</u>	5	6	7	8	9	10
1	1	2	3	4	5	6	7	8	9	10
2	2	4	6	8	10	12	14	16	18	20
3	3	6	9	12	15	18	21	24	27	30
4	4	8	12	16	20	24	28	32	36	40
<u>5</u>	5	10	15	<u>20</u>	25	30	35	40	45	50
6	6	12	18	24	30	36	42	48	54	60
7	7	14	21	28	35	42	49	56	63	70
8	8	16	24	32	40	48	56	64	72	80
9	9	18	27	36	45	54	63	72	81	90
10	10	20	30	40	50	60	70	80	90	100

Grade 7 Chapter 1 Lesson 5
First Way of Multiplication
Example 1

```
Top   3
  405
 ×106
 2430 (P1)
+40500 (P2)
 42930 (P)
```

1) $6 \times 5 = 30$
2) Put 0 below 6 (P1)
3) Put 3 above 0 (top)
4) $6 \times 0 = 0$
5) $0 + 3 = 3$
6) Put 3 ahead 0 (P1)
7) $6 \times 4 = 24$
8) Put 24 ahead 30 (P1)
9) $405 \times 100 = 40500$ (P2)
10) $2430 + 40500 = 42930$ (P)

Example 2

```
Top    1 2
   6304
  × 15
  31520 (P1)
 +63040 (P2)
  94560 (P)
```

1) $5 \times 4 = 20$
2) Put 0 below 5 (P1)
3) Put 2 above 0 (top)
4) $5 \times 0 = 0$
5) $0 + 2 = 2$
6) Put 2 ahead 0 (P1)
7) $5 \times 3 = 15$
8) Put 5 ahead 2 (P1)
9) Put 1 above 6 (top)
10) $5 \times 6 = 30$
11) $30 + 1 = 31$
12) Put 31 ahead 5 (P1)
13) $6304 \times 10 = 63040$ (P2)
14) 94560 (P)

Second Way of Multiplication
Example 3

```
   405
  ×106
  2430 (P1)
+40500 (P2)
 42930 (P)
```

1) $106 = 100 + 6$

2)
```
 405
 × 6
2430 (P1)
```

3)
```
  405
 ×100
40500 (P2)
```

Example 4

```
  6304
  × 15
 31520 (P1)
+63040 (P2)
 94560 (P)
```

1) $15 = 10 + 5$

2)
```
 6304
  × 5
31520 (P1)
```

3)
```
 6304
 × 10
63040 (P2)
```

4) $31520 + 63040 = 94,560$ (Product=P)

Grade 7 Chapter 1 Lesson 5

Review 1 A

To multiply two numbers put in first place number with more digit for easy multiplication. Multiply numbers

1) $\begin{array}{r} 15 \\ \times 4 \\ \hline \end{array}$

2) $\begin{array}{r} 169 \\ \times 6 \\ \hline \end{array}$

3) $\begin{array}{r} 1046 \\ \times 9 \\ \hline \end{array}$

4) $\begin{array}{r} 12008 \\ \times 6 \\ \hline \end{array}$

5) 9×14006

6) $\begin{array}{r} 86 \\ \times 39 \\ \hline \end{array}$

7) $\begin{array}{r} 203 \\ \times 16 \\ \hline \end{array}$

8) $\begin{array}{r} 3006 \\ \times 15 \\ \hline \end{array}$

9) $\begin{array}{r} 304 \\ \times 607 \\ \hline \end{array}$

10) 19×4006

Review 1 B

To multiply two numbers put in first place number with more digit for easy multiplication. Multiply numbers

1) $\begin{array}{r} 16 \\ \times 6 \\ \hline \end{array}$

2) $\begin{array}{r} 178 \\ \times 7 \\ \hline \end{array}$

3) $\begin{array}{r} 1057 \\ \times 8 \\ \hline \end{array}$

4) $\begin{array}{r} 13009 \\ \times 7 \\ \hline \end{array}$

5) 8×15007

6) $\begin{array}{r} 76 \\ \times 48 \\ \hline \end{array}$

7) $\begin{array}{r} 304 \\ \times 17 \\ \hline \end{array}$

8) $\begin{array}{r} 4002 \\ \times 16 \\ \hline \end{array}$

9) $\begin{array}{r} 406 \\ \times 706 \\ \hline \end{array}$

10) 18×5007

Grade 7 Chapter 1 Lesson 6
Division of Whole Numbers

The fourth operation with numbers is division ÷ or :
Division means to separated number by some equal groups

The first number in division is called the dividend	6
The second number in division is called the divisor	2
The result of division is called the quotient	3
Dividend ÷ Divisor = Quotient	$6 \div 2 = 3$

Properties of Division

Any number divided by one is the same number:	$5 \div 1 = 5$
Any number divided by itself is one:	$5 \div 5 = 1$
Zero divided by any number is zero:	$0 \div 5 = 0$
Division by zero is not allowed	$5 \div 0 = \emptyset$

Formulas: 1) $A \div 1 = A$ 2) $A \div A = 1$
3) $0 \div A = 0$ 4) $A \div 0 = \emptyset$ $\emptyset$ - no solutions

To check multiplication use division
$2 \times 3 = 6$ Check $6 \div 2 = 3$ or $6 \div 3 = 2$
To check division use multiplication or division
$6 \div 2 = 3$ Check $2 \times 3 = 6$ or $6 \div 3 = 2$

Division by 10

To divide numbers by 10, delete one zero: $40 \div 10 = 4$

Division by 100

To divide numbers by 100, delete two zeros: $400 \div 100 = 4$

Even numbers 16, 48

Numbers where last digits are even 0, 2, 4, 6, 8 called even numbers

Odd numbers 31, 27

Number where last digits are odd 1, 3, 5, 7, 9, called odd numbers

Average

The quotient of division sum of numbers, by number of addend, called the average of numbers.
Find average of numbers 50, 60, 70, 80, 90
$(50 + 60 + 70 + 80 + 90) \div 5 = 350 \div 5 = 70$
$(10 + 11 + 12 + 13 + 14) \div 5 = 60 \div 5 = 12$

Grade 7 Chapter 1 Lesson 6
Division the Numbers
Exact Division without remainder R = 0

```
    5
7) 35
  -35
    0
```

35 Dividend 7 Divisor 5 Quotient 0 Remainder

Division with remainder R > 0

```
    5
7) 36
  -35
    1
```

Remainder must be less than divisor. If 36 divides by 7, we get:
Dividend - 36, Divisor - 7, Partial Quotient - 5, Remainder – 1
To find the dividend in division with remainder, multiply divisor by the partial quotient and add remainder.
Dividend = Divisor× Partial Quotient + Remainder: 36 = 7 × 5 + 1

Example 1

15280 – dividend 5 – divisor 3056 – quotient

Every time bring down just one digit. Example 1

3056	1) 5 × 3 = 15	Put 3 in quotient
5 ⎸15280	2) 15 – 15 = 0	
-15	3) Bring down the 2	
02	4) 5 × 0 = 0	Put 0 in quotient
-0	5) 2 – 0 = 2	
28	6) Bring down the 8	
-25	7) 5 × 5 = 25	Put 5 in quotient
30	8) 28 – 25 = 3	
- 30	9) Bring down the 0	
0	10) 5 × 6 = 30	Put 6 in quotient
	11) 30 – 30 = 0	

Grade 7 Chapter 1 Lesson 6
Division the Numbers
Example 2

17530 – dividend　　5 – divisor　　3506 – quotient

```
  3506
5|17530
 -15
  25
 -25
   03
   -0
    30
  - 30
```

1) $5 \times \underline{3} = 15$ — Put 3 in quotient
2) $17 - 15 = 2$
3) Bring down the 5
4) $5 \times \underline{5} = 25$ — Put 5 in quotient
5) $25 - 25 = 0$
6) Bring down the 3
7) $5 \times \underline{0} = 0$ — Put 0 in quotient
8) $3 - 0 = 3$
9) Bring down the 0
10) $5 \times \underline{6} = 30$ — Put 6 in quotient
11) $30 - 30 = 0$

Example 3

$6 \div 2 = 3$　　dividend　6　　divisor　2　　quotient　3

Every time bring down just one digit

```
        5070
607) 3071420
    -3035
       364
       -0
       3642
     -3642
         0
```

1) $607 \times 5 = 3035$
2) $3071 - 3035 = 36$
3) Bring down the 4
4) $607 \times \underline{0} = 0$
5) $364 - 0 = 364$
6) Bring down the 2
7) $607 \times \underline{6} = 3642$
8) $3642 - 3642 = 0$
9) Put last 0 from dividend in quotient

Grade 7 Chapter 1 Lesson 6

Example 4

Every time bring down just one digit

```
           607
5060 ) 3071420
      -30360
        3542
         -0
        35420
      - 35420
           0
```

1) 5060 × 6 = 30360
2) 30714 – 30360 = 354
3) Bring down the 2
4) 5060 × 0 = 0
5) 3542 – 0 = 3542
6) Bring down the 0
7) 5060 × 7 = 35420
8) 35420 – 35420 = 0

Example 5

```
   2406 (q)
5 | 12030
   -10
     20
    -20
      03
      -0
       30
      -30
```

1) 5 × 2 = 10 Put 2 in quotient (q)
2) 12 – 10 = 2
3) Bring down the 0
4) 5 × 4 = 20 Put 4 in quotient (q)
5) 20 – 20 = 0
6) Bring down the 3
7) 5 × 0 = 0 Put 0 in quotient (q)
8) 3 – 0 = 3
9) Bring down the 0
10) 5 × 6 = 30 Put 6 in quotient (q)
11) 30–30=0

Review 1 A

Find the average of numbers

1) 60, 70
2) 80, 90
3) 60, 70, 80
4) 70, 80, 90
5) 50, 60, 70, 80
6) 60, 70, 80, 90
7) 55, 60, 65, 70, 75
8) 75, 80, 85, 90, 95
9) 40, 50, 60, 70, 80, 90
10) 50, 55, 60, 60, 65, 70

Grade 7 Chapter 1 Lesson 7

Complete the Table of Division 1 by using Table of Multiplication

$5 \times 4 = 20$ $\quad$ $20 \div 4 = 5$ $\quad$ $20 \div 5 = 4$

Review 1 B

		a	b	c	d	e	f	g	h
	×	2	?	4	?	6	?	8	?
1)	1								9
2)	?								18
3)	3							24	
4)	?							32	
5)	5			20			35		
6)	?					36			
7)	7				35				
8)	?			32					
9)	9		27						
10)	?	20							

Review 2 A Divide the numbers

1) $250 \div 25$
2) $500 \div 5$
3) $750 \div 25$
4) $1000 \div 5$
5) $1250 \div 25$
6) $7500 \div 125$
7) $8750 \div 125$
8) $10000 \div 125$
9) $11250 \div 125$
10) $12500 \div 125$
11) $816 \div 8$
12) $1624 \div 8$
13) $2432 \div 8$
14) $3240 \div 8$
15) $4048 \div 8$
16) $4856 \div 8$
17) $5664 \div 8$
18) $6472 \div 8$
19) $7264 \div 8$
20) $6456 \div 8$

Review 2 B Divide the numbers

1) $16032 \div 16$
2) $32048 \div 16$
3) $48064 \div 16$
4) $64080 \div 16$
5) $80096 \div 16$
6) $96112 \div 16$
7) $112128 \div 16$
8) $128144 \div 16$
9) $144128 \div 16$
10) $128112 \div 16$
11) $32064 \div 32$
12) $64096 \div 32$
13) $96128 \div 32$
14) $128160 \div 32$
15) $160192 \div 32$
16) $192224 \div 32$
17) $224256 \div 32$
18) $256192 \div 32$
19) $288256 \div 32$
20) $256224 \div 32$

Grade 7 Chapter 1 Lesson 7
Four Operations with numbers

1) Multiplication numbers: for memory <u>(x, x, any)</u>
To multiply numbers, work in any order
$9\times8\ \times 7 = 72 \times 7 = 504$ $\qquad$ $9 \times 8 \times 7 = 9 \times 56 = 504$

2) Work from left to right: for memory <u>(L → R)</u>
a)Multiplication and Division numbers: (× or ÷)
To multiply and divide numbers work from left to right (L → R)
$9\times6 \div 3 = 54 \div 3 = 18$ $\qquad$ $4 \times 81 \div 9 = 324 \div 9 = 36$
b)Division and Multiplication numbers: (÷ or ×)
To divide and multiply numbers work from left to right <u>(L → R)</u>
$24\div6 \times 2 = 4 \times 2 = 8$ $\qquad$ $81 \div 9 \times 4 = 9 \times 4 = 36$
c)Division numbers: (÷, ÷)
To divide numbers, work from left to right <u>(L → R)</u>
$24\div4 \div 2 = 6 \div 2 = 3$ $\qquad$ $810\div 9\div 3 = 90\div 9 = 10$

3) Do Multiplication or Division before Addition or Subtraction for memory (× or ÷) before (+ or -)
a)Multiplication and Addition numbers: 1) × 2) +
To multiply and add numbers always first multiply than add
$9\times 6 + 9 = 54 + 9 = 63$ $\qquad$ $9 + 9 \times 9 = 9 + 81 = 90$
b)Multiplication and Subtraction numbers: 1) × 2) −
To multiply and subtract numbers always first multiply than subtract $9\times9 - 9 = 81 - 9 = 72$ $\qquad$ $90 - 9 \times 9 = 90 - 81 = 9$
c)Division and Addition numbers: 1) ÷ 2) +
To divide and add numbers always first divide than add
$50\div 5 + 39 = 10 + 39 = 49$ $\qquad$ $88 + 171 \div 9 = 88 + 19 = 107$
d)Division and Subtraction numbers: 1) ÷ 2) −
To divide and subtract numbers always first divide than subtract
$108 \div 6 - 9 = 18 - 9 = 9$ $\qquad$ $500 - 126 \div 7 = 500 - 18 = 482$

Grade 7 Chapter 1 Lesson 8
Reviews for Test 1
Review 1 A

Read directions carefully. **Answers must be in the indicated place.**
Show all steps for full credits. **Do only 1 operation per 1 step.**
Solve 1 problem 5 minutes. **Each problem is worth 5 points.**

Evaluate

1) $\begin{array}{r} 900000 \\ \underline{-\ 456789} \end{array}$

1)______

2) $\begin{array}{r} 987 \\ \underline{\times 806} \end{array}$

2)______

3) $\begin{array}{r} 908 \\ \underline{\times 605} \end{array}$

3)______

4) $437658 \div 543$

4)______

5) $304920 \div 504$

5)______

Grade 7 Chapter 1 Lesson 8
Review 1 A

Read directions carefully. Answers must be in the indicated place.
Show all steps for full credits. Do only 1 operation per 1 step.
Solve 1 problem 5 minutes. Each problem is worth 5 points.

Evaluate

6) $300300 \div 60 \times 5$

6)______

7) $9250 - 650 \times 5$

7)______

8) $90081 - 81072 \div 9$

8)______

9) $(50 \times 300) \div 6$

9)______

10) $361 \div (117 - 98)$

10)______

Grade 7 Chapter 1 Lesson 8
Review 1 B

Read directions carefully. Answers must be in the indicated place.
Show all steps for full credits. Do only 1 operation per 1 step.
Solve 1 problem 5 minutes. Each problem is worth 5 points.

Calculate

1) $\begin{array}{r} 800000 \\ \underline{-\ 345678} \end{array}$

1)______

2) $\begin{array}{r} 876 \\ \underline{\times 705} \end{array}$

2)______

3) $\begin{array}{r} 807 \\ \underline{\times 706} \end{array}$

3)______

4) $304560 \div 432$

4)______

5) $284518 \div 403$

5)______

Grade 7 Chapter 1 Lesson 8
Review 1 B

Read directions carefully. **Answers must be in the indicated place.**
Show all steps for full credits. **Do only 1 operation per 1 step.**
Solve 1 problem 5 minutes. **Each problem is worth 5 points.**

Evaluate

6) $280280 \div 70 \times 4$

6)______

7) $8960 - 740 \times 4$

7)______

8) $72064 - 64056 \div 8$

8)______

9) $(40 \times 420) \div 7$

9)______

10) $324 \div (105 - 87)$

10)______

Grade 7 Chapter 1 Lesson 9
Tests for Chapter 1
Test 1 A

Read directions carefully. Answers must be in the indicated place.
Show all steps for full credits. Do only 1 operation per 1 step.
Solve 1 problem 5 minutes. Each problem is worth 5 points.

Evaluate

1) $\begin{array}{r} 700000 \\ \underline{-\ 234567} \end{array}$

1)______

2) $\begin{array}{r} 765 \\ \underline{\times 604} \end{array}$

2)______

3) $\begin{array}{r} 706 \\ \underline{\times 605} \end{array}$

3)______

4) $193884 \div 321$

4)______

5) $243714 \div 302$

5)______

Grade 7 Chapter 1 Lesson 9
Test 1 A

Read directions carefully. Answers must be in the indicated place.
Show all steps for full credits. Do only 1 operation per 1 step.
Solve 1 problem 5 minutes. Each problem is worth 5 points.

Evaluate

6) $240240 \div 80 \times 3$

6)______

7) $7490 - 830 \times 3$

7)______

8) $56049 - 49042 \div 7$

8)______

9) $(30 \times 560) \div 8$

9)______

10) $289 \div (93 - 76)$

10)______

Grade 7 Chapter 1 Lesson 9

Test 1 B

Read directions carefully. Answers must be in the indicated place.
Show all steps for full credits. Do only 1 operation per 1 step.
Solve 1 problem 5 minutes. Each problem is worth 5 points.

Evaluate

1) $\begin{array}{r} 600000 \\ \underline{-\ 123456} \end{array}$

1)______

2) $\begin{array}{r} 654 \\ \underline{\times 503} \end{array}$

2)______

3) $\begin{array}{r} 605 \\ \underline{\times 403} \end{array}$

3)________

4) $105630 \div 210$

4)________

5) $182508 \div 201$

5)_________

Grade 7 Chapter 1 Lesson 9
Test 1 B

Read directions carefully. Answers must be in the indicated place.
Show all steps for full credits. Do only 1 operation per 1 step.
Solve 1 problem 5 minutes. Each problem is worth 5 points.

Evaluate

6) $180180 \div 90 \times 2$

6)______

7) $6840 - 920 \times 2$

7)______

8) $42036 - 36030 \div 6$

8)______

9) $(20 \times 720) \div 9$

9)______

10) $256 \div (81 - 65)$

10)______

Grade 7 Chapter 1 Lesson 10

Summary 5

Finding unknown

Rules and Examples

Addition Numbers

First number in addition called the addend 2

Second number in addition called the addend 3

Result of addition called the sum 5

Addend + Addend = Sum $2 + 3 = 5$

First Addend = Sum – Second Addend
? + 3 = 5 ? = 5 – 3 ? = 2 5 > 2

Second Addend = Sum – First Addend
2 + ? = 5 ? = 5 – 2 ? = 3 5 > 3

Subtraction Numbers

First number in subtraction called the minuend 5

Second number in subtraction called the subtrahend 2

Result of multiplication called the product 3

Minuend – Subtrahend = Difference $5 - 2 = 3$

Minuend = Difference + Subtrahend
? - 2 = 3 ? = 3 + 2 ? = 5 5 > 3

Subtrahend = Minuend – Difference
5 - ? = 3 ? = 5 – 3 ? = 2 5 > 2

Grade 7 Chapter 1 Lesson 10
Summary 5
Finding unknown
Rules and Examples

Multiplication Numbers

First number in multiplication called the factor 2

Second number in multiplication called the factor 3

Result of multiplication called the product 6

Factor × Factor = Product $2\times 3 = 6$

First Factor = Product ÷ Second Factor

$? \times 3 = 6$ $? = 6 \div 3$ $? = 2$ $6 > 2$

Second Factor = Product ÷ First Factor

$2 \times ? = 6$ $? = 6 \div 2$ $? = 3$ $6 > 3$

Division Numbers

First number in division is called dividend 6

Second number in division is called divisor 2

Result of division is called quotient 3

Dividend ÷ Divisor = Quotient $6 \div 2 = 3$

Divisor = Dividend ÷ Quotient

$6 \div ? = 3$ $? = 6 \div 3$ $? = 2$ $6 > 2$

Dividend = Quotient × Divisor

$? \div 2 = 3$ $? = 3\times 2$ $? = 6$ $6 > 3$

Grade 7 Chapter 1 Lesson 10

Review 1 A

Add

In column # 1 19190 + 11910 # 2 28280 + 12820

Fund the unknowns
3 ? + 37370 = 51100
4 46460 +? = 61100
5 59590 + 15950 = ?

Complete the table

6) ?+12920=42210

	# 6	# 7	# 8	# 9	# 10
Addend	?	38380	47470		68680
Addend	12920		14740	15650	
Sum	42210	52210		72210	85540

Subtract

In column # 11 31100 - 19190 # 12 41100 - 28280

Find the unknowns
13 ? - 37370 = 13730
14 61100 - ? = 14640
15 75540 – 59590 = ?

Complete the table

16)?-29290=12920

	# 16	# 17	# 18	# 19	# 20
Minuend	?	52210	62210		85540
Subtrahend	29290		47470	56560	
Difference	12920	13830		15650	16860

Grade 7 Chapter 1 Lesson 10

Review 1 B

Add

In column # 1 39390 + 13930 # 2 48480 + 14840

Find the unknowns
3 ? + 15750 = 73320
4 69690 +? = 86650
5 79790 + 17970 =?

Complete the table

10)89890+?=98870

	# 6	# 7	# 8	# 9	# 10
Addend		58580	67670		89890
Addend	14940		16760	17870	?
Sum	64430	74430		96650	98870

Subtract

In column # 11 53320 - 39390 # 12 63320 - 48480

Find the unknowns
13 ? - 57570 = 15750
14 86650 –? = 16960
15 97760 – 79790 =?

Complete the table

20)98870-?=8980

	# 16	# 17	# 18	# 19	# 20
Minuend		74430	84430		98870
Subtrahend	49490		67670	78780	?
Difference	14940	15850		17870	8980

Grade 7 Chapter 1 Lesson 10

Review 2 A
Find unknown number

1 $1 \times ? = 5006$
2 $? \times 1 = 6007$
3 $2 \times ? = 14016$
4 $? \times 3 = 24027$
5 $4 \times 9009 = ?$

	# 6	# 7	# 8	# 9	# 10
Factor	?	65	?	85	95
Factor	100	?	75	?	405
Product	5500	6630	15225	25840	?

11 $? \div 9 = 9008$
12 $64056 \div ? = 8$
13 $? \div 7 = 7006$
14 $36030 \div ? = 6$
15 $25020 \div 5 = ?$

	# 16	# 17	# 18	# 19	# 20
Dividend	?	9045	?	28035	40500
Divisor	10	?	25	?	45
Quotient	5040	15	702	35	?

Grade 7 Chapter 1 Lesson 10
Review 2 B
Find unknown number

1 $5 \times ? = 5000$
2 $? \times 6 = 6012$
3 $7 \times ? = 14021$
4 $? \times 8 = 24032$
5 $9 \times 4005 = ?$

	# 6	# 7	# 8	# 9	# 10
Factor	?	15	?	35	909
Factor	10	?	25	?	45
Product	5060	9105	17700	28315	?

11 $16012 \div ? = 4$
12 $? \div 3002 = 3$
13 $4002 \div ? = 2$
14 $? \div 2003 = 1$
15 $1002 \div 1 = ?$

	# 16	# 17	# 18	# 19	# 20
Dividend	?	6630	?	25840	38475
Divisor	55	?	75	?	95
Quotient	100	65	203	85	?

Grade 7 Answers Chapter 1

Lesson 1 Pretest 1

1) 406728 2) 609 3) 12012 4) 240240 5) 27

Lesson 1 College Placement Test B

1) 29	2) 70	3) 1.2	4) - 3.5	5) - 2c – 3 d
6) 2	7) 12 ft	8) 35 in^2	9) 2.5 hr.	10) 12 girl

Lesson 2 Review 1

A	1) 427,130	2) 807	3) 9,009	4) 260,260	5) 35
B	1) 243,815	2) 908	3) 4,004	4) 180,180	5) 24

Lesson 3 Review 1 A

1) 120	2) 230	3) 350	4) 460	5) 570
6) 680	7) 790	8) 900	9) 990	10) 880
11) 1,200	12) 2,300	13) 3,500	14) 4,600	15) 5,700
16) 6,800	17) 7,900	18) 9,000	19) 9,900	20) 8,800

Lesson 3 Review 1 B

1) 1< 14	2) 345 > 23	3) 456 < 1,234
4) 12,345 > 2,345	5) 2,345 < 12,345	6) 23,456 > 9,876
7) 3,456 < 23,456	8) 34,567 > 8,765	9) 9,876 = 9,876
10) 8,765 = 8,765	11) 9,876 > 9,080	12) 1,234 < 1,325
13) 1,001 = 1,001	14) 1,234 < 1,543	15) 9,876 > 9,767
16) 1,234 < 1,243	17) 3,456 < 3,465	18) 3,465 > 3,456
19) 1,243 > 1,234	20) 2,345 < 2,354	

Lesson 4 Review 1 A

	a	b	c	d	e
1)	311,000	411,000	511,000	611,000	755,400
2)	422,100	522,100	622,100	722,100	855,400
3)	533,200	633,200	733,200	866,500	977, 600
4)	644,300	744,300	844,300	966,500	988,700
5)	119,100	128,200	137,300	146,400	159, 500
6)	129,200	138,300	147,400	156,500	168,600
7)	139,300	148,400	157,500	169,600	179,700
8)	149,400	158,500	167,600	178,700	89,800

Grade 7 Answers Chapter 1

Lesson 4 Review 1 B

	a	b	c	d	e
1)	3,110,000	4,110,000	5,110,000	6,110,000	7,554,000
2)	4,221,000	5,221,000	6,221,000	7,221,000	8,554,000
3)	5,332,000	6,332,000	7,332,000	8,665,000	9,776,000
4)	6,443,000	7,443,000	8,443,000	9,665,000	9,887,000
5)	1,191,000	1,282,000	1,373,000	1,464,000	1,595,000
6)	1,292,000	1,383,000	1,474,000	1,565,000	1,686,000
7)	1,393,000	1,484,000	1,575,000	1,696,000	1,797,000
8)	1,494,000	1,585,000	1,676,000	1,787,000	898,000

Lesson 5 Review 1 A

1) 60 2) 1,014 3) 9,414 4) 72,048 5) 126,054
6) 3,354 7) 3,248 8) 45,090 9) 184,528 10) 76,114

Lesson 5 Review 1 B

1) 96 2) 1,246 3) 8,456 4) 91,063 5) 120,056
6) 3,648 7) 5,168 8) 64,032 9) 286,636 10) 90,126

Lesson 6 Review 1 A

1) 65 2) 85 3) 70 4) 80 5) 65

6) 75 7) 65 8) 85 9) 65 10) 60

Lesson 7 Review 2 A

1) 10 2) 100 3) 30 4) 200 5) 50
6) 60 7) 70 8) 80 9) 90 10) 100
11) 102 12) 203 13) 304 14) 405 15) 506
16) 607 17) 708 18) 809 19) 908 20) 807

Lesson 7 Review 2 B

1) 1,002 2) 2,003 3) 3,004 4) 4,005 5) 5,006
6) 6,007 7) 7,008 8) 8,009 9) 9,008 10) 8,007
11) 1,002 12) 2,003 13) 3,004 14) 4,005 15) 5,006
16) 6,007 17) 7,008 18) 8,006 19) 9,008 20) 8,007

Grade 7 Answers Chapter 1

Lesson 8 Review 1 A

1) 443211 2) 795522 3) 549340 4) 806 5) 605

6) 25025 7) 6000 8) 81073 9) 2500 10) 19

Lesson 8 Review 1 B

1) 454322 2) 617580 3) 569742 4) 705 5) 706

6) 16016 7) 6000 8) 64057 9) 2400 10) 18

Lesson 9 Test 1 A

1) 465433 2) 462060 3) 427130 4) 604 5) 807

6) 9009 7) 5000 8) 49043 9) 2100 10) 17

Lesson 9 Test 1 B

1) 476544 2) 328962 3) 243815 4) 503 5) 908

6) 4004 7) 5000 8) 36031 9) 1600 10) 16

Grade 7 Answers Chapter 1

Lesson 10 Review 1 A

#1	31100	#2	41100	#3	13730	#4	14640	#5	75540
#6	29290	#7	13830	#8	62210	#9	56560	#10	16860
#11	11910	#12	12820	#13	51100	#14	46460	#15	15950
#16	42210	#17	38380	#18	14740	#19	72210	#20	68680

Lesson 10 Review 1 B

#1	53320	#2	63320	#3	57570	#4	16960	#5	97760
#6	49490	#7	15850	#8	84430	#9	78780	#10	8980
#11	13930	#12	14840	#13	73320	#14	69690	#15	17970
#16	64430	#17	58580	#18	16760	#19	96650	#20	89890

Lesson 10 Review 2 A

# 1	5006	# 6	55	# 11	81072	# 16	50400
# 2	6007	# 7	102	# 12	8007	# 17	603
# 3	7008	# 8	203	# 13	49042	# 18	17550
# 4	8009	# 9	304	# 14	6005	# 19	801
# 5	36036	# 10	38475	# 15	5004	# 20	900

Lesson 10 Review 2 B

# 1	1000	# 6	506	# 11	4003	# 16	5500
# 2	1002	# 7	607	# 12	9006	# 17	102
# 3	2003	# 8	708	# 13	2001	# 18	15225
# 4	3004	# 9	809	# 14	2003	# 19	304
# 5	36045	# 10	40905	# 15	1002	# 20	405

Grade 7

Chapter 2

Properties of Whole Numbers

Grade 7 Chapter 2 Contents
Properties of whole numbers
Lesson 1 – 10
Basic Problems

Work with lesson 1, read rules and examples, 2) Solve the problems from contents, 3) Check answers in the end of chapter 1, 4) Work with the next lessons 2, 3, 4, 5, …, 5) Solve Test 1 A, 6) Grading Scale: A =10 points, B=9points, C=8 points, D=7 points, F=0-6 points, 7) If you solve 10 problems correctly, you get an A, 8) If have an A, you can move on to next chapter, 9) If your grade is lower than an A, work with the mistakes, then solve Test 1B, 10) When you finish all of the chapters, solve Final Exam in last chapter.

Lesson 1 Memorize table of multiplication. Solve the Pretest 1 #1–#5 Lesson 2 Solve Mental Math # 1 - # 4 all without paper and pen, mentally. Lesson 3 Power of Whole numbers. Memorize Table 2 Perfect Square Numbers. Prime Factorization. Solve Review 1A # 36-# 40, Review 1 B # 36 – # 40 Lesson 4 Tests for division. Solve Review 1 A # 1- # 10, Review 1 B # 1-# 10

Lesson 5 Greatest Common Factor (GCF), Least Common Multiple (LCM) Solve Review 1 A # 16 – # 20. Solve Review 1 B # 16 – # 20
Lesson 6 Order of five operations
Solve Review 1 A # 36 – # 40. Solve Review 1 B # 36 – # 40
Lesson 7 Laws for whole numbers
Solve the Review 1 A # 21 – # 30, Solve the Review 1 B # 21 – # 30
Lesson 8 Common factor
Solve Review 1 A # 6 – # 10, Review 1 B # 6 – # 10
Solve Review 1 C # 6 – # 10, Review 1 D # 6 – # 10

Lesson 9 Reviews for Test 2
Solve the Review 2 A #1–# 10 all. Solve the Review 2 B #1–# 10 all
Lesson 10 Tests 2 Solve Test 2 A #1–# 10 all. Solve Test 2 B # 1 – # 10 all
Answers Chapter 2

Grade 7 Chapter 2 Lesson 1
Table 1
Multiplication and Division

A + 0 = A	0 + A = A	A – 0 = A	A – A = 0
A × 1 = A	1 × A = A	A ÷ 1 = A	A ÷ A = 1
A × 0 = 0	0 × A = 0	0 ÷ A = 0	A ÷ 0 = Ø

A + B = B + A   A × B = B × A   Ø - no solutions

Number 5 from the left column multiply by number 4 from first row to get answer 20. 5 × 4 = 20 20 ÷ 5 = 4 20 ÷ 4 = 5

Memorize Table 1

×	1	2	3	<u>4</u>	5	6	7	8	9	10
1	1	2	3	4	5	6	7	8	9	10
2	2	4	6	8	10	12	14	16	18	20
3	3	6	9	12	15	18	21	24	27	30
4	4	8	12	16	20	24	28	32	36	40
<u>5</u>	5	10	15	<u>20</u>	25	30	35	40	45	50
6	6	12	18	24	30	36	42	48	54	60
7	7	14	21	28	35	42	49	56	63	70
8	8	16	24	32	40	48	56	64	72	80
9	9	18	27	36	45	54	63	72	81	90
10	10	20	30	40	50	60	70	80	90	100

Grade 7 Chapter 2 Lesson 1
Pretest
Evaluate

1) $8 \cdot 225 - 8 \cdot 75$

1)___

2) $12912 \div 16 - 8064 \div 16$

2) _____

3) $3^4 - 4^3 \div 2^3$

3) _____

4) $70200 \div (95^2 - 35^2)$

4) ______

5) $\{3 \cdot 8^2 - 2^2[2^3 + (1 + 3)^2]\} + 2^3$

5) _____

Mental Math 1

Add

1) 9 + 1
2) 8 + 2
3) 7 + 3
4) 6 + 4
5) 5 + 5
6) 9 + 2
7) 8 + 3
8) 7 + 4
9) 6 + 5
10) 9 + 3
11) 8 + 4
12) 7 + 5
13) 6 + 6
14) 9 + 4
15) 8 + 5
16) 7 + 6
17) 9 + 5
18) 8 + 6
19) 7 + 7
20) 9 + 6
21) 8 + 7
22) 9 + 7
23) 8 + 8
24) 9 + 8
25) 9 + 9

Mental Math 2
Subtract

1) 10 – 1
2) 10 – 2
3) 10 – 3
4) 10 – 4
5) 10 – 5
6) 11 - 2
7) 11 – 3
8) 11 – 4
9) 11 – 5
10) 12 – 3
11) 12 – 4
12) 12 – 5
13) 12 – 6
14) 13 – 4
15) 13 – 5
16) 13 – 6
17) 14 – 5
18) 14 – 6
19) 14 – 7
20) 15 – 6
21) 15 – 7
22) 16 – 7
23) 16 – 8
24) 17 – 8
25) 18 - 9

Grade 7 Chapter 2 Lesson 2

Mental Math 3

Multiply

1) 5×6
2) 6×6
3) 7×6
4) 8×6
5) 9×6
6) 5×7
7) 6×7
8) 7×7
9) 8×7
10) 9×7
11) 5×8
12) 6×8
13) 7×8
14) 8×8
15) 9×8
16) 5×9
17) 6×9
18) 7×9
19) 8×9
20) 9×9
21) 5×10
22) 6×10
23) 7×10
24) 8×10
25) 9×10

Mental Math 4
Divide

1) 30 ÷ 6
2) 36 ÷ 6
3) 42 ÷ 6
4) 48 ÷ 6
5) 54 ÷ 6
6) 35 ÷ 7
7) 42 ÷ 7
8) 49 ÷ 7
9) 56 ÷ 7
10) 63 ÷ 7
11) 40 ÷ 8
12) 48 ÷ 8
13) 56 ÷ 8
14) 64 ÷ 8
15) 72 ÷ 8
16) 45 ÷ 9
17) 54 ÷ 9
18) 63 ÷ 9
19) 72 ÷ 9
20) 81 ÷ 9
21) 50 ÷ 10
22) 60 ÷ 10
23) 70 ÷ 10
24) 80 ÷ 10
25) 90 ÷ 10

Grade 7 Chapter 2 Lesson 3
Power of Whole Numbers.

The four arithmetic operations with numbers are:
Addition, Subtraction, Multiplication, Division
$3 + 2 = 5$ $5 - 2 = 3$ $3 \times 2 = 6$ $6 \div 2 = 3$

The fifth operation with numbers is power of numbers.
A power of a number indicates repeated multiplication.

Base is a number that has to be multiplied 3
Exponent is a number that shows how many times
to multiply the base 2
Power is result of multiplication of the same numbers 9

$3^2 = 3 \times 3 = 9$ **$\text{Base}^{\text{Exponent}} = \text{Power}$**

Exponent 2 called squared **$4^2 = 4 \times 4 = 16$**
Exponent 3 called cubed **$2^3 = 2\cdot2\cdot2 = 4\cdot2 = 8$**

Special powers

$0^2 = 0$	$0^{10} = 0$	$0^n = 0$
$1^2 = 1$	$1^{10} = 1$	$1^n = 1$

Perfect square numbers

$1^2 = 1$	$2^2 = 4$	$3^2 = 9$	$4^2 = 16$	$5^2 = 25$
$6^2 = 36$	$7^2 = 49$	$8^2 = 64$	$9^2 = 81$	$10^2 = 100$

Perfect square numbers are 1, 4, 9, 16, 25, 36, 49, 64, 81, 100

$10^2 + 11^2 + 12^2 = 13^2 + 14^2$ $100 + 121 + 144 = 169 + 196$ $365 = 365$

$3^4 = 3\times3\times3\times3 = 81$	$3^5 = 81 \times 3 = 243$	$3^6 = 243 \times 3 = 729$
$4^4 = 4\times4\times4\times4 = 256$	$4^5 = 256 \times 4 = 1024$	$5^4 = 125 \times 5 = 625$

Grade 7 Chapter 2 Lesson 3

Table 2
Perfect Square Numbers

Take number 20 from the left column add number 5 from the first row to get number 25
Answer is in intersection row of 20 and column of 5
$25^2 = 25 \cdot 25 = 625$

Memorize Table of Perfect Square for Indicated Numbers

a^2	0	1	2	3	4	5	6	7	8	9
10	100	121	144	169	196	225	256	289	324	361
20	400	441	484	529	576	625	676	729	784	841
30	900	961	1024	1089	1156	1225	1296	1369	1444	1521
40	1600	1681	1764	1849	1936	2025	2116	2209	2304	2401
50	2500	2601	2704	2809	2916	3025	3136	3249	3364	3481
60	3600	3721	3844	3969	4096	4225	4356	4489	4624	4761
70	4900	5041	5184	5329	5476	5625	5776	5929	6084	6241
80	6400	6561	6724	6889	7056	7225	7396	7569	7744	7921
90	8100	8281	8464	8649	8836	9025	9216	9409	9604	9801

Grade 7 Chapter 2 Lesson 3

Prime Factorization

Factor × Factor = Product **$3 \times 2 = 6$**
A factor of a number n is a nonzero whole number that divides evenly into n.

All divisors of 12 are 1, 2, 3, 4, 6, 12 **$1\times12 = 2\times6 = 3\times4 = 12$**

All divisors of 24 are 1, 2, 3, 4, 6, 8, 12, 24 **$1\times24=2\times12=3\times8=4\times6=24$**

Prime Numbers

A natural number called a prime number if it has only two different divisors: one and the number itself **2, 3, 5, 7, 11, 13, 17, 19 …**

Composite Numbers

A natural number that has more than two divisors called a composite number 4, 6, 8, 9, 10, 12, 14, 15, 16…
The number 1 has only one divisor: which is why 1 is not a prime or a composite. Prime numbers cannot factorize.

Remember the first 10 prime numbers between 1 and 30
2, 3, 5, 7, 11, 13, 17, 19, 23, 29
To factorize composite numbers to prime factors means to represent composite numbers in the product of prime factors:
$8 = 4 \times 2 = 2 \times 2 \times 2 = 2^3$

Prime Factorization

Example 1: Find prime factorization for 100
Try to divide by prime numbers: by 2, by 3, by 5, by 7…
$100 \div 2$
$50 \div 2$
$25 \div 5$
$5 \div 5$
$100 = 2\times2\times5\times5 = 2^2\cdot5^2$
Example 2:
Represent your numbers like the product of just prime numbers
$50 = 2 \times 25 = 2 \times 5 \times 5 = 2\times 5^2$

Grade 7 Chapter 2 Lesson 3

Review 1 A

Find the powers of numbers

1)	11^2	2)	12^2	3)	13^2	4)	14^2	5)	15^2
6)	16^2	7)	17^2	8)	18^2	9)	19^2	10)	20^2
11)	21^2	12)	22^2	13)	23^2	14)	24^2	15)	25^2
16)	26^2	17)	27^2	18)	28^2	19)	29^2	20)	31^2
21)	5^2	22)	15^2	23)	25^2	24)	35^2	25)	45^2
26)	55^2	27)	65^2	28)	75^2	29)	85^2	30)	95^2
31)	1^3	32)	2^3	33)	3^3	34)	4^3	35)	5^3
36)	6^3	37)	7^3	38)	8^3	39)	9^3	40)	10^3

Review 1 B

Find Prime Factorization of Numbers

1)	4	2)	6	3)	8	4)	9	5)	10
6)	12	7)	14	8)	15	9)	16	10)	18
11)	20	12)	21	13)	24	14)	25	15)	27
16)	28	17)	30	18)	32	19)	35	20)	36
21)	40	22)	42	23)	45	24)	48	25)	49
26)	50	27)	54	28)	56	29)	60	30)	63
31)	64	32)	70	33)	72	34)	75	35)	80
36)	81	37)	84	38)	90	39)	96	40)	120

Grade 7 Chapter 2 Lesson 4

Tests for division

Test for division is a way to know, without doing division, if some other number can divide a number evenly or with a remainder of zero. A number is divisible by another number when there is no remainder.

Test for 2

A number is divisible by 2 if its last digit is even digits (0, 2, 4, 6, 8)
If a number can divide by 2 this number called even.
Example: 1,234

Test for 5

A number is divisible by 5 if its last digit is 0 or 5
Example: 100 can be divided by 5 $100 \div 5 = 20$

Test for 10

A number is divisible by 10 if its last digits have one or more zeros
Example: 240 can be divided by 10 $240 \div 10 = 24$

Test for 25

A number is divisible by 25 if the last two digits to the right are 00, 25, 50, 75. Example: 100, 250, 675, 925 $100 \div 25 = 4$

Test for 4

A number is divisible by 4 if the last two digits to the right of the number are 00 or make a new number that can be divided by 4.
Example:1$\underline{00}$, 7$\underline{28}$, 5$\underline{40}$ $100 \div 4 = 25$ $28 \div 4 = 7$ $40 \div 4 = 10$

Test for 8

A number is divisible by 8 if the last three digits to the right of number are 000 or make a new number that can be divided by 8.
Example: 1$\underline{000}$, 2$\underline{160}$, 6$\underline{240}$
$1000 \div 8 = 125$ $160 \div 4 = 40$ $240 \div 4 = 60$

Grade 7 Chapter 2 Lesson 4

Test for 3

A number is divisible by 3 if the sum of its digits is divisible by 3
$342 = 300 + 40 + 2 = 3\times100 + 4\times10 + 2 = 3(99+1) + 4(9+1) + 2$
$3 \times 99+3+4\times9+4+2 = 3\times99+4\times9 + (3+4+2)$
Example: 342 $3+4+2 = 9$ $9\div3 = 3342\div3 = 114$

Test for 9

A number is divisible by 9 if the sum of its digits is divisible by 9.
$468 = 400 + 60 + 8 = 4\times100+6\times10+8 = 4(99+1) + 6(9+1) + 8$
$4 \times 99+ 4 + 6\times 9 + 6 + 8 = 4\times99 + 6\times9 + (4 + 6 + 8)$
Example:468 $4+6+8 = 18$ $18\div9 = 2$ $468 \div9 = 52$

Test for 6

A number is divisible by 6 if it is divisible by both 2 and 3.
Example: 246 Last digit 6 $2+4+6=12$ $12\div3 = 4$ $246\div6=41$

Test for 15

A number is divisible by 15 if it is divisible by both 5 and 3.
Example: 465 Last digit 5 $4 + 6 + 5 = 15$ $15\div3 = 5$ $465\div15 = 31$

Test for Sum

If each addend can divide by that number the sum divisible by a number $(a + b) \div c = a \div c + b \div c$
$(18 + 12) \div 3 = 18 \div 3 + 12 \div 3 = 6 + 4 = 10$

Test for Difference

If minuend and subtrahend can divide by a number the difference divisible by a number $(a - b) \div c = a \div c - b \div c$
$(18-12)\div3=18\div3-12\div3=6-4=2$

Test for Product

If one of factors divide by number the product divisible by a number
$(A \times B) \div C = (A \div C) \times B$ $(6 \times 5) \div 3 = (6 \div 3) \times 5 = 2 \times 5 = 10$
$(A \times B) \div C = A \times (B \div C)$ $(7 \times 6) \div 3 = 7 \times (6 \div 3) = 7 \times 2 = 14$

Grade 7 Chapter 2 Lesson 4

Mental Math means: if you can do operations mentally

If numbers can divide evenly, try to multiply or divide easily

1)To multiply a number by 5, divide this number by 2, and multiply that result by 10. $16\times5 = 16\div2\times10 = 8\times10 = 80$.

$24\times5 = 24\div2\times10 = 12\times10 = 120$

2)To multiply a number by 25, divide this number by 4, and multiply that result by 100. $16\times25=16\div4\times100=4\times100=400$.

$24\times25=24\div4\times100=6\times100=600$

3)To multiply a number by 125, divide this number by 8, and multiply result by 1000. $16\times125=16\div8\times1000=2\times1000=2000$.

$24\times125=24\div8\times1000=3\times1000=3000$

4)To divide a number by 5, divide this number by 10, and multiply that result by 2. $50\div5=50\div10\times2=5\times 2=10$.

$80\div5=80\div10\times2=8\times2=16$

5)To divide a number by 25, divide this number by 100, and multiply that result by 4. $500\div25=500\div100\times4=5\times4=20$.

$800\div25=800\div100\times4=8\times4=32$

6)To divide a number by 125, divide this number by 1000, and multiply that result by 8. $5000\div125=5000\div1000\times8=5\times8=40$.

$8000\div125=8000\div1000\times8=8\times8=64$

Review 1 A Prove and check:

Can given numbers be divided or not by some numbers?

1)	Given Numbers	120, 133, 224, 241, 346	by 2
2)	Given Numbers	100, 186, 245, 362, 465	by 5
3)	Given Numbers	140, 275, 350, 567, 460	by 10
4)	Given Numbers	150, 165, 225, 320, 475	by 25
5)	Given Numbers	140, 186, 244, 326, 480	by 4
6)	Given Numbers	160, 280, 2160, 1006, 2320	by 8
7)	Given Numbers	180, 190, 240, 466, 393	by 3
8)	Given Numbers	180, 284, 468, 682, 567	by 9
9)	Given Numbers	186, 273, 684, 328, 534	by 6
10)	Given Numbers	165, 235, 735, 234, 645	by 15

Review 1 B Evaluate the expressions

1)	$(180 + 120) \div 3$	2)	$(180 - 120) \div 3$
3)	$(180 + 120) \div 5$	4)	$(180 - 120) \div 5$
5)	$(60 \times 5) \div 3$	6)	$(5 \times 60) \div 3$
7)	$(90 \times 7) \div 3$	8)	$(50 \times 7) \div 5$
9)	$(7 \times 50) \div 5$	10)	$(80 \times 9) \div 5$

Grade 7 Chapter 2 Lesson 5

Divisors and Multiples
Greatest Common Factor

The product of the numbers 2 and 5 is 10; the numbers 2 and 5 called factors. If the number 10 divided by 2, the result is 5 and the divisor is 2. The number 2 can called a factor and can called a divisor of 10.

Greatest Common Factor or Greatest Common Divisor
(in short GCF or GCD)

1) Find the GCD of numbers 12 and 18

Divisors of number 12 are 1, 2, 3, 4, 6, 12

Divisors of number 18 are 1, 2, 3, 6, 9, 18

Common divisors are 1, 2, 3, 6

Greatest common divisor is 6

GCD (12, 18) = 6 $\quad 12 \div 6 = 2 \quad 18 \div 6 = 3$

2) Find GCF of numbers 24, 36, and 48

Factors of number 24 are 1, 2, 3, 4, 6, 8, 12, 24

Factors of number 36 are 1, 2, 3, 4, 6, 12, 18, 36

Factors of number 48 are 1, 2, 3, 4, 6, 12, 16, 24, 48

Common factors are 1, 2, 3, 4, 6, 12

Greatest common factor is 12

GCF (24, 36, 48) = 12 $\quad 24 \div 12=2 \quad 36\div12=3 \quad 48\div12=4$

3) If the greatest number divided evenly by the smallest number, the greatest common factor of numbers is equal to the smallest of given numbers: GCF (12, 4) = 4

Factors of 12 are 1, 2, 3, 4, 6, 12

Factors of 4 are 1, 2, 4

Common factors are 1, 2, 4

Greatest common factor is 4

GCF (12, 4) = 4 $\quad 12 \div 4 = 3$

4) The greatest common factor of prime numbers is one:

GCF (5,7) = 1 $\quad 5 \div 1 = 5 \quad 7\div1 = 7$

Grade 7 Chapter 2 Lesson 5

Multiples

Any numbers, which is evenly divided by a given number, called the Multiple of a given number: $10 \div 5 = 2$ 10 is a multiple of 5

Any number has many multiples. Multiples of 5 are 5, 10, 15, 20 …

The least multiple for any number is the number itself.

Least Common Multiple, or in short LCM (× 1, 2, 3, 4, 5…)

The least number is the smallest number that can evenly divide by each of the given numbers.

1) Find the LCM of numbers 12 and 18

Multiples of 12 are 12, 24, 36, 48, 60, 72…

Multiples of 18 are 18, 36, 54, 72…

Common multiples are 36, 72…

Least common multiples is 36

LCM (12, 18) = 36 $36 \div 12 = 3$ $36 \div 18 = 2$

2)If the greatest of given numbers evenly divided by the smallest numbers,

the greatest number is the least common multiple of the given numbers. Find the LCM of numbers 4, 12

Multiples of 4 are 4, 8, 12, 16, 20, 24 …

Multiples of 12 are 12, 24 …

Common multiples are 12, 24 …

LCM (4, 12) = 12

3)The least common multiple of prime numbers is equal to the product of the numbers.

Find the LCM of numbers 2 and 5

Multiples of 2 are 2, 4, 6, 8, 10…

Multiples of 5 are 5, 10…

Common multiples are 10, 20 …

Least multiple is 10

LCM (2, 5) = $2 \times 5 = 10$

4) A second way to find LCM. Find the LCM of numbers 12 and 18

If the greatest number cannot be divided by the smallest number, try to multiply the big number by 2, 3, 4, 5 … and check division

Example: $18 > 12$ 18 cannot be divided evenly by 12

$18 \times 2 = 36$ $36 \div 12 = 3$ LCM (12, 18) = 36

Grade 7 Chapter 2 Lesson 5

Review 1 A

Find Greatest Common Factor of Given Numbers (GCF)

1)	(6, 9)	2)	(8, 12)	3)	(12, 18)	4)	(18, 24)
5)	(20, 25)	6)	(21, 28)	7)	(20, 30)	8)	(16, 24)
9)	(18, 27)	10)	(36, 24)	11)	(24, 40)	12)	(15, 45)
13)	(24, 8)	14)	(32, 48)	15)	(36, 48)	16)	(12, 24)
17)	(9, 18, 36)	18)	(8, 16, 32)	19)	(6, 12, 24)	20)	(12, 24, 36)

Review 1 B

Find Least Common Multiply of Given Numbers (LCM)

1)	(4, 10)	2)	(6, 8)	3)	(6, 10)	4)	(12, 8)
5)	(12, 18)	6)	(16, 20)	7)	(18, 24)	8)	(25, 20)
9)	(28, 21)	10)	(20, 30)	11)	(16, 24)	12)	(27, 18)
13)	(12, 24)	14)	(15, 45)	15)	(15, 20)	16)	(14, 28)
17)	(12, 8, 6)	18)	(6, 9, 18)	19)	(6, 4, 12)	20)	(12, 16, 24)

Grade 7 Chapter 2 Lesson 6

Order of Five Operations

An arithmetic expression contains numbers, sign of operations and sometimes parentheses.

Result of all operations called value of expressions.

Expressions with numbers can be in different form:

as a sum 8 + 4, a difference 8 - 4, a product 8 × 4, or a quotient 8 ÷ 4.

Operations of the first level are addition (+) and subtraction (-)

Operations of the second level are multiplication (×) and division (÷)

Operations of the third level are power (a^n), root, and logarithm

Order of Operations of the Same Level (+, -) or (×, ÷)

An arithmetic expression without parenthesis, that has just addition or just multiplication, add or multiply in any order:

80 + 60 + 20 = 80 + 20 + 60 = 160   25 × 6 × 4 = 25 × 4 × 6 = 600

11 + 22 + 19 + 28 = (11 + 19) + (22 + 28) = 30 + 50 = 80
4 × 5 × 25 × 2 = (4 × 25) × (5 × 2) = 100 × 10 = 1000

An expression without parenthesis that has operations with the same level of operations (+, -) or (×, ÷), work in order, from left to the right.

80 – 20 + 40 = 60 + 40 = 100   160 ÷ 4 × 2 = 40 × 2 = 80
38 – 8 + 32 – 2 = 30 + 32 – 2 = 62 – 2 = 60
64 ÷ 8 × 4 ÷ 2 = 8 × 4 ÷ 2 = 32 ÷ 2 = 16

Grade 7 Chapter 2 Lesson 6

Order of Operations of Two Levels (+, -, ×, ÷)

To work an expression without parentheses that has operations of two levels of operations, at first work in order with the second level of operation (×, ÷) from left to right, then work in order with the first level of operations (+, -,) from left to right

$60 + 40 \times 2 = 60 + 80 = 140$

$180 \div 9 + 40 = 20 + 40 = 60$

$20 \times 4 - 20 = 80 - 20 = 60$

$80 - 150 \div 3 = 80 - 50 = 30$

$12 + 18 \times 2 \div 2 = 12 + 36 \div 2 = 12 + 18 = 30$
$18 - 24 \div 6 \times 2 = 18 - 4 \times 2 = 18 - 8 = 10$

$16 \times 4 \div 2 - 6 = 64 \div 2 - 6 = 32 - 6 = 26$
$18 \div 9 \times 2 + 4 = 2 \times 2 + 4 = 4 + 4 = 8$

Order of Operations in an Expression with Parentheses

In an expression with parentheses: the first step is to work with the operations in parentheses, and the second step is to work before and after parentheses from left to right:

$(50 + 40) \times 2 = 90 \times 2 = 180$

$3 \times (90 - 70) = 3 \times 20 = 60$

$(180 - 40) \div 7 = 140 \div 7 = 20$

$200 \div (40 - 20) = 200 \div 20 = 10$

$15 \div 3 \times (5 + 4) \times 2 = 15 \div 3 \times 9 \times 2 = 5 \times 9 \times 2 = 45 \times 2 = 90$
$20 \div (4 - 2) \times 3 = 20 \div 2 \times 3 = 10 \times 3 = 30$

$6 \times (9 - 7) \div 3 = 6 \times 2 \div 3 = 12 \div 3 = 4$
$(18 - 4) \div 7 \times 5 = 14 \div 7 \times 5 = 2 \times 5 = 10$

Grade 7 Chapter 2 Lesson 6

Grouping Symbols

Expressions have three kinds of grouping symbols: first do the operations in parentheses, second do the operations - in brackets, and third do the operations - in braces:

Parentheses ( ) **Brackets [ ]** **Braces { }**

Order of Five Operations

Expressions have five operations: first do the operations of the third level (a^n); second, do the operations of the second level (× or ÷), and lastly, do the operations of the first level (+ or -), in order as they occur from left to right.

Remember (in short):

Grouping Symbols	**(1) [2] {3}**	
Exponents (Power)		**a^n**
Multiplication or Division from left to right	**× or ÷**	**L → R**
Addition or Subtract from left to right	**+ or –**	**L → R**

Example 1

$6^2 \div 2^2 \times 3 - 3^2 \times 6 \div 3 = 36 \div 4 \times 3 - 9 \times 6 \div 3 = 9 \times 3 - 54 \div 3 = 27 - 18 = 9$

Example 2

$50 - \{200 \div [60 - (30 + 20)]\} = 50 - \{200 \div [60 - 50]\} =$
$50 - 200 \div 10 = 50 - 20 = 30$

Grade 7 Chapter 2 Lesson 6
Review 1 A
Evaluate the expressions

1) $3 \times 5 + 40 \div 8$
2) $30 \div 5 + 8 \times 9$
3) $8 \times 5 - 20 \div 5$
4) $84 \div 4 - 8 \times 2$

5) $8 \times 11 + 90 \div 10$
6) $13 \times 6 + 96 \div 12$
7) $12 \times 7 - 88 \div 11$
8) $14 \times 7 - 91 \div 13$

9) $15 \times 6 + 90 \div 15$
10) $17 \times 4 + 85 \div 17$
11) $16 \times 5 - 96 \div 16$
12) $18 \times 3 - 90 \div 18$

13) $185 \div 5 + 215 \div 5$
14) $136 \div 4 + 104 \div 4$
15) $546 \div 7 - 336 \div 7$
16) $558 \div 6 - 138 \div 6$

17) $456 \div 8 + 264 \div 8$
18) $846 \div 9 + 396 \div 9$
19) $12912 \div 16 - 11296 \div 16$
20) $14528 \div 16 - 9680 \div 16$

21) $(60 - 20) \times 2 \div 4$
22) $(50 + 40) \times 2 \div 3$
23) $5 \times (10 + 40) \div 5$
24) $3 \times (90 - 70) \div 5$

25) $(80 - 30) \times 2 \div 25$
26) $200 \div (40 - 20) \times 5$
27) $(60 + 40) \div 10 \times 2$
28) $(180 - 40) \div 7 \times 20$

29) $300 \div (13 + 17) \times 24$
30) $(54 \times 4 - 16) \div 5$
31) $(500 - 250) \div 50 \times 16$
32) $15 \times (48 + 32) \div 16$

33) $(400 - 100) \times 9 \div 100$
34) $2400 \div (35 + 65) \times 10$
35) $324 \div (64 - 46) \times 5$
36) $(90 \times 5 - 50) \div 20$

37) $45 \div 5 \times (6 + 4) \times 2$
38) $60 \div 5 \times (16 + 14) \times 2$
39) $80 \div 8 \times (24 \div 6 \times 4)$
40) $60 \div 6 \times (48 \div 6 \times 4)$

Grade 7 Chapter 2 Lesson 6

Review 1 B

Perform the operations

1) $2^4 \div 2^2 - 2$

2) $3^4 \div 3^2 - 3$

3) $4^3 \div 2^2 - 2^3$

4) $6^3 \div 2^2 - 3^3$

5) $8^3 \div 2^2 - 4^3$

6) $9^3 \div 3^2 - 3^3$

7) $3^4 - 3^3 \div 3^2$

8) $4^3 - 4^2 \div 2^2$

9) $6^3 - 6^2 \div 3^2$

10) $8^2 - 6^2 \div 2^2$

11) $2^2 - 8^2 \div 4^2$

12) $4^2 - 4^2 \div 2^2$

13) $5^2 - 6^2 \div 2^2$

14) $6^2 - 8^2 \div 4^2$

15) $7^2 - 9^2 \div 3^2$

16) $8^2 - 6^2 \div 3^2$

17) $2^3 \div (4 - 2)^2$

18) $3^3 \div (6 - 3)^2$

19) $4^3 \div (8 - 4)^2$

20) $5^3 \div (9 - 4)^2$

Grade 7 Chapter 2 Lesson 6

Review 1 B
Perform the operations

21) $6^3 \div (5 - 2)^2$

22) $8^3 \div (7 - 3)^2$

23) $2^5 - 2^4 \div 2^2$

24) $2^5 - 2^4 \div 2^3$

25) $2^6 - 2^5 \div 2^4$

26) $2^6 - 2^5 \div 2^3$

27) $2^8 - 2^7 \div 2^6$

28) $2^9 - 2^8 \div 2^7$

29) $2^8 - 2^9 \div 2^7$

30) $4^3 \div 2^3 - 1^3$

31) $3^4 - 3^4 \div 3^2$

32) $3^5 - 3^4 \div 3^3$

33) $3^6 - 3^5 \div 3^4$

34) $3^4 - 4^3 \div 2^3$

35) $6^2 - 5^2 + 4^2$

36) $5^4 - 4^4 - 3^4$

37) $6^3 - 5^3 - 3^3$

38) $9^2 - 8^2 + 7^2$

39) $9^3 - 8^3 + 7^3$

40) $7^2 - 6^2 + 5^2$

Grade 7 Chapter 2 Lesson 7

Laws for Whole Numbers

A formula is a statement of math relations between variables. Formulas include letters that can represent any numbers. Equality is two expressions joined by the equal symbol.

Laws of Addition

First Law of Addition (commutative)

Changing the order of two addends does not change the sum:

$A + B = B + A$ $2 + 3 = 5$ $3 + 2 = 5$ $2 + 3 = 3 + 2$

Second Law of Addition (associative)

Changing the grouping of addends does not change the sum:

$A + (B + C) = (A + B) + C$ $8 + (2 + 5) = (8 + 2) + 5 = 10 + 5 = 15$

Use the commutative law of addition as the first step.
Use the associative law of addition as the second step.
Add in any order. Use the two laws together.

$2 + 7 + 3 + 8 = (2 + 8) + (3 + 7) = 10 + 10 = 20$

Laws of Multiplication

First Law of Multiplication (commutative)

Changing the order of two factors does not change the product:

$A B = B A$ $2 \times 3 = 6$ $3 \times 2 = 6$ $2 \times 3 = 3 \times 2$

Second Law of Multiplication (associative)

Changing the grouping of factors does not change the product:

$A (B C) = (AB) C$ $5 \times (2 \times 4) = (5\times2) \times 4 = 10 \times 4 = 40$

Use the commutative law of multiplication as the first step
Use the associative law of multiplication as the second step
Multiply in any order; use the two laws together:

$5 \times 8 \times 2 = 5 \times 2 \times 8 = 10 \times 8 = 80$

$2\times5 = 10$ $2\times50 = 100$ $4\times25 = 100$ $8\times125 = 1000$

Grade 7 Chapter 2 Lesson 7

First Distributive Law of Multiplication with Addition

To multiply a number by a sum, multiply that number by each of the addends and add the results

A (B+C) =AB+AC 2(3+5) = 2× 3+ 2×5= 6 +10=16 2(3+5)=2×8=16

(A + B) C = AC+BC   32 × 5= (30+2)×5 = 30×5+2×5 =150+10=160

Second Distributive Law of Multiplication with Subtraction

To multiply a number by a difference, multiply that number by the minuend and subtrahend, and subtract the results:

A (B – C) = AB – AC   2(10 – 4) = 2 × 10 – 2 × 4 = 20 – 8 = 12

(A - B) C = AC – B C   (20-2)×5 =20×5–2×5=100–10=90

Division sum   A÷C + B÷C = (A + B) ÷ C
18 ÷ 3 + 12 ÷ 3 = (18 + 12) ÷ 3 = 30 ÷ 3 = 10

Division difference   A ÷ C – B ÷ C = (A - B) ÷ C
18 ÷ 3 – 12 ÷ 3 = (18 - 12) ÷ 3 = 6 ÷ 3 = 2

To do mental calculations means to carry out operations by laws or properties.
24×11=24(10+1)=24×10+24×1=240+24=264 (2+4=6 -middle digit)

Memorize Formulas

1) **A + (B + C) = (A + B) + C**
2) **A × (B × C) = (A × B) × C**
3) **A · (B + C) = A·B + A·C**
4) **A · (B - C) = A·B - A·C**
5) **A·B + A·C = A · (B + C)**
6) **A·B - A·C = A · (B - C)**
7) **A ÷ C + B ÷ C = (A + B) ÷ C**
8) **A ÷ C - B ÷ C = (A - B) ÷ C**

Grade 7 Chapter 2 Lesson 7
Review 1 A
Add numbers

$10 = 1 + 9 = 2 + 8 = 3 + 7 = 4 + 6 = 5 + 5$

1) $43 + 18 + 27$
2) $22 + 46 + 18$
3) $16 + 59 + 24$
4) $53 + 22 + 17$
5) $15 + 99 + 85$
6) $81 + 46 + 19$
7) $88 + 39 + 12$
8) $36 + 47 + 53$
9) $21 + 35 + 79$
10) $86 + 20 + 14$
11) $35 + 65 + 16$
12) $11 + 89 + 15$
13) $(95 + 90) + 5$
14) $(10 + 88) + 12$
15) $(33 + 56) + 67$
16) $(19 + 14) + 86$
17) $(92 + 3) + 8$
18) $3 + (97 + 18)$
19) $(24 + 23) + 76$
20) $(25 + 17) + 75$

Multiply numbers

$100 = 2 \times 50 = 4 \times 25 = 5 \times 20 = 10 \times 10$

21) $5 \times 25 \times 2$
22) $25 \times 5 \times 4$
23) $(20 \times 9) \times 5$
24) $(50 \times 6) \times 2$
25) $(25 \times 6) \times 4$
26) $(25 \times 7) \times 4$
27) $25 \times (8 \times 4)$
28) $25 \times (4 \times 9)$
29) $8 \times (5 \times 20)$
30) $20 \times (4 \times 5)$

Grade 7 Chapter 2 Lesson 7
Review 1 B
Evaluate by using laws

1) $5 \cdot (10 + 6)$
2) $5 \cdot (30 + 8)$
3) $(20 + 7) \times 5$
4) $(40 + 9) \times 5$
5) $25 \times (10 + 1)$
6) $25 \times (10 - 4)$
7) $8\,(20 - 5)$
8) $8\,(30 + 5)$
9) $(10 - 2) \times 25$
10) $(20 - 4) \times 25$

Evaluate by using laws

$$34 \times 5 = (30 + 4) \times 5 = 30 \times 5 + 4 \times 5 = 150 + 20 = 170$$

11) 12×5
12) 14×5
13) 16×5
14) 18×5
15) 22×5
16) 12×25
17) 14×25
18) 16×25
19) 18×25
20) 22×25

Perform the operations

21) $150 \div 5 + 200 \div 5$
22) $(200 + 35) \div 5$
23) $160 \div 4 + 120 \div 4$
24) $(360 + 12) \div 6$
25) $480 \div 8 - 240 \div 8$
26) $(100 - 15) \div 5$
27) $810 \div 9 - 360 \div 9$
28) $(400 - 16) \div 8$
29) $560 \div 7 - 350 \div 7$
30) $(540 - 120) \div 6$

Grade 7 Chapter 2 Lesson 8
Common Factor 1

A formula is a statement of math relations between variables. Formulas include letters that can represent any numbers. Equality is two expressions joined by the equal symbol.

Laws of Addition

First Law of Addition (commutative)

$A + B = B + A$ $\quad 2 + 3 = 5 \quad 3 + 2 = 5 \quad 2 + 3 = 3 + 2$

Second Law of Addition (associative)

$A + (B + C) = (A + B) + C \quad 8 + (2 + 5) = (8 + 2) + 5 = 10 + 5 = 15$

Add in any order. Use the two laws together.

$2 + 7 + 3 + 8 = (2 + 8) + (3 + 7) = 10 + 10 = 20$

Law of Multiplication

First Law of Multiplication (commutative)

$A\ B = B\ A \quad 2 \times 3 = 6 \quad 3 \times 2 = 6 \quad 2 \times 3 = 3 \times 2$

Second Law of Multiplication (associative)

$A\ (B\ C) = (AB)\ C \quad 5 \times (2 \times 4) = (5\times2) \times 4 = 10 \times 4 = 40$

Multiply in any order; use the two laws together:

$5 \times 8 \times 2 = 5 \times 2 \times 8 = 10 \times 8 = 80$

First Distributive Law of Multiplication with Addition

$A\ (B + C) = AB + AC \quad 2(3+5) = 2\times 3+ 2\times5= 6 +10=16$

$(A + B)\ C = AC + BC$

$14 \times 5= (10 + 4) \times 5 = 10 \times 5 + 4 \times 5 =50+20=70$

Second Distributive Law of Multiplication with Subtraction

$A\ (B – C) = AB – AC \quad 2(10 – 4) = 2 \times 10 – 2 \times 4 = 20 – 8 = 12$

$(A - B)\ C = AC – B\ C \quad (20 - 2) \times 5=20\times5–2\times5=100–10=90$

To do mental calculations means to carry out operations by laws or properties.

24×11=24(10+1)=24×10+24×1=240+24=264 (2+4=6 -middle digit)

Grade 7 Chapter 2 Lesson 8
Common Factor 1
Memorize Formulas

1) $A \cdot (B + C) = A \cdot B + A \cdot C$
2) $A \cdot (B - C) = A \cdot B - A \cdot C$
3) $A \cdot B + A \cdot C = A \cdot (B + C)$
4) $A \cdot B - A \cdot C = A \cdot (B - C)$
5) $34 \cdot 5 = (30 + 4) \cdot 5 = 30 \cdot 5 + 4 \cdot 5 = 150 + 20 = 170$
6) $8 \cdot 5 + 2 \cdot 5 = 5 \cdot (8 + 2) = 5 \cdot 10 = 50$

Review 1 A

1) $80 \times 7 + 20 \times 7$
2) $70 \times 5 - 20 \times 5$
3) $70 \times 4 + 30 \times 4$
4) $60 \times 6 - 40 \times 6$

5) $40 \times 5 + 60 \times 5$
6) $90 \times 4 - 80 \times 4$
7) $50 \times 20 + 50 \times 80$
8) $30 \times 7 - 30 \times 5$

9) $30 \times 40 + 30 \times 60$
10) $50 \times 90 - 50 \times 70$

Review 1 B

1) $80 \times 75 + 80 \times 25$
2) $25 \times 60 - 15 \times 60$
3) $20 \times 17 + 20 \times 13$
4) $75 \times 40 - 35 \times 40$

5) $30 \times 27 + 30 \times 23$
6) $93 \times 20 - 20 \times 53$
7) $14 \times 40 + 40 \times 16$
8) $70 \times 28 - 70 \times 18$

9) $29 \times 30 + 31 \times 30$
10) $77 \times 20 - 20 \times 47$

Grade 7 Chapter 2 Lesson 8
Common Factor 2
Memorize Formulas

1) $A \cdot (B + C) = A \cdot B + A \cdot C$
2) $A \cdot (B - C) = A \cdot B - A \cdot C$
3) $A \cdot B + A \cdot C = A \cdot (B + C)$
4) $A \cdot B - A \cdot C = A \cdot (B - C)$

$A\,(B + C) = AB + AC$
$5 \times (40 + 10) = 5 \times 40 + 5 \times 10 = 200 + 50 = 250$

$AB + AC = A\,(B + C)$
$40 \times 5 + 10 \times 5 = 5\,(40 + 10) = 5(50) = 250$

$A\,(B - C) = AB - AC$
$5\,(40 - 10) = 5 \times 40 - 5 \times 10 = 200 - 50 = 150$

$AB - AC = A\,(B - C)$
$5 \times 40 - 5 \times 10 = 5\,(40 - 10) = 5(30) = 150$

$AB + AC = A\,(B + C)$ $\qquad$ $AB - AC = A\,(B - C)$
If two products have a common factor, put the common factor before parentheses. AB + AC A – common factor Put A before parentheses

In parentheses, put results of division products by common factor
$AB \div A = B$ $\qquad$ $AC \div A = C$

$16 \cdot 5 + 4 \cdot 5 = 5\,(16 + 4) = 5\,(20) = 100$
$22 \cdot 5 - 2 \cdot 5 = 5\,(22 - 2) = 5\,(20) = 100$

$75 \times 5 + 25 \times 5 = 5(75 + 25) = 5(100) = 500$
$125\,(5) - 25(5) = 5(125 - 25) = 5(100) = 500$

$36 \cdot 5 = (30 + 6) \cdot 5 = 30 \cdot 5 + 6 \cdot 5 = 150 + 30 = 180$
$8 \cdot 9 + 2 \cdot 9 = 9 \cdot (8 + 2) = 9 \cdot 10 = 90$

Grade 7 Chapter 2 Lesson 8
Common Factor 2
Memorize Formulas

$A \cdot (B + C) = A \cdot B + A \cdot C$

$A \cdot B + A \cdot C = A \cdot (B + C)$

$A \cdot (B - C) = A \cdot B - A \cdot C$

$A \cdot B - A \cdot C = A \cdot (B - C)$

Review 1 C

1) $80\times75 + 80\times25$
2) $37\times40 + 40\times13$

3) $25\times40 - 15\times40$
4) $54\times50 - 34\times50$

5) $70\times37 + 63\times70$
6) $49\times40 + 40\times51$

7) $85\times60 - 25\times60$
8) $39\times80 - 80\times19$

9) $30\times27 + 30\times23$
10) $24\times30 + 16\times30$

Review 1 D

1) $93 \times 20 - 20 \times 63$
2) $58 \times 30 + 42 \times 30$

3) $43 \times 40 + 40 \times 17$
4) $96 \times 20 - 26 \times 20$

5) $50 \times 28 - 50 \times 18$
6) $38 \times 45 - 18 \times 45$

7) $29 \times 30 + 31 \times 30$
8) $30 \times 60 + 70 \times 60$

9) $77 \times 40 - 40 \times 47$
10) $80 \times 57 - 80 \times 47$

Grade 7 Chapter 2 Lesson 9
Reviews for Test 2
Review 2 A

Read directions carefully. **Answers must be in the indicated place.**
Show all steps for full credits. **Do only 1 operation per 1 step.**
Solve 1 problem 5 minutes. **Each problem is worth 5 points.**

Evaluate with laws with whole numbers

1) $(A + B) + C = (A + C) + B$
$(85 + 80) + 5$

2) $(A \cdot B) \cdot C = (A \cdot C) \cdot B$
$(20 \cdot 58) \cdot 5$

3) $A\,(B + C) = A \cdot B + A \cdot C$
$4 \cdot (50 + 25)$

4) $A\,(B - C) = A \cdot B - A \cdot C$
$4 \cdot (50 - 25)$

5) $A \cdot B + A \cdot C = A\,(B + C)$
$8 \cdot 125 + 8 \cdot 25$

6) $A \cdot B - A \cdot C = A\,(B - C)$
$8 \cdot 125 - 8 \cdot 25$

7) $22592 \div 32 - 19360 \div 32$

Evaluate

8) $2^3 - 8^2 \div 4^2$

9) $58800 \div (95^2 - 25^2)$

10) $6^3 - 2^3(10 - 2^3 \div 2^2)$

Grade 7 Chapter 2 Lesson 9
Review 2 B

Read directions carefully. **Answers must be in the indicated place.**
Show all steps for full credits. **Do only 1 operation per 1 step.**
Solve 1 problem 5 minutes. **Each problem is worth 5 points.**

Evaluate by using laws with whole numbers

1) $(A + B) + C = (A + C) + B$
$(80 + 88) + 12$

2) $(A \cdot B) \cdot C = (A \cdot C) \cdot B$
$(50 \cdot 67) \cdot 2$

3) $A \cdot (B + C) = A \cdot B + A \cdot C$
$8 \cdot (125 + 50)$

4) $A \cdot (B - C) = A \cdot B - A \cdot C$
$8 \cdot (125 - 50)$

5) $A \cdot B + A \cdot C = A \cdot (B + C)$
$4 \cdot 75 + 4 \cdot 50$

6) $A \cdot B - A \cdot C = A \cdot (B - C)$
$4 \cdot 75 - 4 \cdot 50$

7) $14528 \div 16 - 9680 \div 16$

Evaluate

8) $6^3 - 6^2 \div 2^2$

9) $52800 \div (85^2 - 25^2)$

10) $\{3 \cdot 10^2 - 2^2[2^2 + (2 + 3)^2]\} + 2^3$

Grade 7 Chapter 2 Lesson 10
Tests for Chapter 2
Test 2 A

Read directions carefully. Answers must be in the indicated place. Show all steps for full credits. Do only 1 operation per 1 step. Solve 1 problem 5 minutes. Each problem is worth 5 points.

Evaluate by using laws with whole numbers

1) $(75 + 200) + 25$

1) ______

2) $(5 \cdot 84) \cdot 20$

2) ______

3) $8 \cdot (125 + 60)$

3) ______

4) $4 \cdot (125 - 50)$

4) ______

5) $8 \cdot 125 + 8 \cdot 75$

5) ______

6) $4 \cdot 175 - 4 \cdot 50$

6) ______

Grade 7 Chapter 2 Lesson 10
Test 2 A

Read directions carefully. **Answers must be in the indicated place.**
Show all steps for full credits. **Do only 1 operation per 1 step.**
Solve 1 problem 5 minutes. **Each problem is worth 5 points.**

Evaluate

7) $29056 \div 32 - 25824 \div 32$

7)______

8) $3^4 - 3^4 \div 3^2$

8) ______

9) $39600 \div (85^2 - 25^2)$

9)______

10) $7^3 - 2^2(3^2 \times 2^3 - 2^2)$

10) _____

Grade 7 Chapter 2 Lesson 10
Test 2 B

Read directions carefully. Answers must be in the indicated place.
Show all steps for full credits. Do only 1 operation per 1 step.
Solve 1 problem 5 minutes. Each problem is worth 5 points.

Evaluate by using laws with whole numbers

1) $(175 + 300) + 25$

1)______

2) $(2 \cdot 96) \cdot 50$

2)______

3) $4 \cdot (125 + 25)$

3)______

4) $8 \cdot (125 - 70)$

4)_____

5) $4 \cdot 150 + 4 \cdot 25$

5)_____

6) $8 \cdot 225 - 8 \cdot 75$

6)___

Grade 7 Chapter 2 Lesson 10

Test 2 B

Read directions carefully. Answers must be in the indicated place.
Show all steps for full credits. Do only 1 operation per 1 step.
Solve 1 problem 5 minutes. Each problem is worth 5 points.

Evaluate

7) $12912 \div 16 - 8064 \div 16$

7) _____

8) $3^4 - 4^3 \div 2^3$

8) _____

9) $70200 \div (95^2 - 35^2)$

9) ______

10) $\{3 \cdot 8^2 - 2^2[2^3 + (1 + 3)^2]\} + 2^3$

10) _____

Grade 7 Answers Chapter 2

Lesson 1 Pretest

1) 1200 2) 303 3) 73 4) 9 5) 104

Lesson 2 Mental Math

Mental Math 1

1) 10	2) 10	3) 10	4) 10	5) 10
6) 11	7) 11	8) 11	9) 11	10) 12
11) 12	12) 12	13) 12	14) 13	15) 13
16) 13	17) 14	18) 14	19) 14	20) 15
21) 15	22) 16	23) 16	24) 17	25) 18

Mental Math 2

1) 9	2) 8	3) 7	4) 6	5) 5
6) 9	7) 8	8) 7	9) 6	10) 9
11) 8	12) 7	13) 6	14) 9	15) 8
16) 7	17) 9	18) 8	19) 7	20) 9
21) 8	22) 9	23) 8	24) 9	25) 9

Mental Math 3

1) 30	2) 36	3) 42	4) 48	5) 54
6) 35	7) 42	8) 49	9) 56	10) 63
11) 40	12) 48	13) 56	14) 64	15) 72
16) 45	17) 54	18) 63	19) 72	20) 81
21) 50	22) 60	23) 70	24) 80	25) 90

Mental Math 4

1) 5	2) 6	3) 7	4) 8	5) 9
6) 5	7) 6	8) 7	9) 8	10) 9
11) 5	12) 6	13) 7	14) 8	15) 9
16) 5	17) 6	18) 7	19) 8	20) 9
21) 5	22) 6	23) 7	24) 8	25) 9

Grade 7 Answers Chapter 2

Lesson 3 Review 1 A

1)	121	2)	144	3)	169	4)	196	5)	225
6)	256	7)	289	8)	324	9)	361	10)	400
11)	441	12)	484	13)	529	14)	576	15)	625
16)	676	17)	729	18)	784	19)	841	20)	961
21)	25	22)	225	23)	625	24)	1225	25)	2025
26)	3025	27)	4225	28)	5625	29)	7225	30)	9025
31)	1	32)	8	33)	27	34)	64	35)	125
36)	216	37)	343	38)	512	39)	729	40)	1000

Lesson 3 Review 1 B

1)	2^2	2)	$2 \cdot 3$	3)	2^3	4)	3^2	5)	$2 \cdot 5$
6)	$2^2 \cdot 3$	7)	$2 \cdot 7$	8)	$3 \cdot 5$	9)	2^4	10)	$2 \cdot 3^2$
11)	$2^2 \cdot 5$	12)	$3 \cdot 7$	13)	$2^3 \cdot 3$	14)	5^2	15)	3^3
16)	$2^2 \cdot 7$	17)	$2 \cdot 3 \cdot 5$	18)	2^5	19)	$5 \cdot 7$	20)	$2^2 \cdot 3^2$
21)	$2^3 \cdot 5$	22)	$2 \cdot 3 \cdot 7$	23)	$3^2 \cdot 5$	24)	$2^4 \cdot 3$	25)	7^2
26)	$2 \cdot 5^2$	27)	$2 \cdot 3^3$	28)	$2^3 \cdot 7$	29)	$2^2 \cdot 3 \cdot 5$	30)	$3^2 \cdot 7$
31)	2^6	32)	$2 \cdot 5 \cdot 7$	33)	$2^3 \cdot 3^2$	34)	$3 \cdot 5^2$	35)	$2^4 \cdot 5$
36)	3^4	37)	$2^2 \cdot 3 \cdot 7$	38)	$2 \cdot 3^2 \cdot 5$	39)	$2^5 \cdot 3$	40)	$2^3 \cdot 3 \cdot 5$

Lesson 4 Review 1 A

1)	Yes, No, Yes, No, Yes	2)	Yes, No, Yes, No, Yes
3)	Yes, No, Yes, No, Yes	4)	Yes, No, Yes, No, Yes
5)	Yes, No, Yes, No, Yes	6)	Yes, No, Yes, No, Yes
7)	Yes, No, Yes, No, Yes	8)	Yes, No, Yes, No, Yes
9)	Yes, No, Yes, No, Yes	10)	Yes, No, Yes, No, Yes

Review 1 B

1)	100	2)	20	3)	60	4)	12	5)	100
6)	100	7)	210	8)	70	9)	70	10)	144

Lesson 5 Review 1 A

1)	3	2)	4	3)	6	4)	6	5)	5
6)	7	7)	10	8)	8	9)	9	10)	12
11)	8	12)	15	13)	8	14)	16	15)	12
16)	12	17)	9	18)	8	19)	6	20)	12

Review 1 B

1)	20	2)	24	3)	30	4)	24	5)	36
6)	80	7)	72	8)	100	9)	84	10)	60
11)	48	12)	54	13)	24	14)	45	15)	60
16)	28	17)	24	18)	18	19)	12	20)	48

Grade 7 Answers Chapter 2

Lesson 6 Review 1 A

1)	20	2)	78	3)	36	4)	5	5)	97
6)	86	7)	76	8)	91	9)	96	10)	73
11)	74	12)	49	13)	80	14)	60	15)	30
16)	70	17)	90	18)	138	19)	101	20)	303
21)	20	22)	60	23)	50	24)	12	25)	4
26)	50	27)	20	28)	400	29)	240	30)	40
31)	80	32)	75	33)	27	34)	240	35)	90
36)	20	37)	180	38)	720	39)	160	40)	320

Lesson 6 Review 1 B

1)	2	2)	6	3)	8	4)	27	5)	64
6)	54	7)	78	8)	60	9)	212	10)	55
11)	0	12)	12	13)	16	14)	32	15)	40
16)	60	17)	2	18)	3	19)	4	20)	5
21)	24	22)	32	23)	28	24)	30	25)	62
26)	60	27)	254	28)	510	29)	252	30)	7
31)	72	32)	240	33)	726	34)	73	35)	27
36)	288	37)	64	38)	66	39)	560	40)	38

Lesson 7 Review 1 A

1)	88	2)	86	3)	99	4)	92	5)	199
6)	146	7)	139	8)	136	9)	135	10)	120
11)	116	12)	115	13)	190	14)	110	15)	156
16)	119	17)	103	18)	118	19)	123	20)	117
21)	250	22)	500	23)	900	24)	600	25)	600
26)	700	27)	800	28)	900	29)	800	30)	400

Lesson 7 Review 1 B

1)	80	2)	190	3)	135	4)	245	5)	275
6)	150	7)	120	8)	280	9)	200	10)	400
11)	60	12)	70	13)	80	14)	90	15)	110
16)	300	17)	350	18)	400	19)	450	20)	550
21)	70	22)	47	23)	70	24)	62	25)	30
26)	17	27)	50	28)	48	29)	30	30)	70

Grade 7 Answers Chapter 2

Lesson 8 Review 1 A

1)	700	2)	250	3)	400	4)	120	5)	500
6)	40	7)	5000	8)	60	9)	3000	10)	1000

Lesson 8 Review 1 B

1)	8000	2)	600	3)	600	4)	1600	5)	1500
6)	800	7)	1200	8)	700	9)	1800	10)	600

Lesson 8 Review 1 C

1)	8000	2)	2000	3)	400	4)	1000	5)	7000
6)	4000	7)	3600	8)	1600	9)	1500	10)	1200

Lesson 8 Review 1 D

1)	600	2)	3000	3)	2400	4)	1400	5)	500
6)	900	7)	1800	8)	6000	9)	1200	10)	800

Lesson 9 Review 2 A

1)	170	2)	5800	3)	300	4)	100	5)	1200
6)	800	7)	101	8)	4	9)	7	10)	152

Lesson 9 Review 2 B

1)	180	2)	6700	3)	1400	4)	600	5)	500
6)	100	7)	303	8)	207	9)	8	10)	192

Lesson 10 Test 2 A

1)	300	2)	8400	3)	1480	4)	300	5)	1600
6)	500	7)	101	8)	72	9)	6	10)	71

Lesson 10 Test 2 B

1)	500	2)	9600	3)	600	4)	440	5)	700
6)	1200	7)	303	8)	73	9)	9	10)	104

Grade 7

Chapter 3

Decimals

Grade 7 Chapter 3 Contents
Decimals
Lesson 1 – 10
Basic Problems

Work with lesson 1, read rules and examples, 2) Solve the problems from contents, 3) Check answers in the end of chapter 1, 4) Work with the next lessons 2, 3, 4, 5, …, 5) Solve Test 1 A, 6) Grading Scale: A =10 points, B=9points, C=8 points, D=7 points, F=0-6 points, 7) If you solve 10 problems correctly, you get an A, 8) If have an A, you can move on to next chapter, 9)If your grade is lower than an A, work with the mistakes, then solve Test 1B,
10) When you finish all of the chapters, solve Final Exam in last chapter.

Lesson 1 Memorize Table 1 of Multiplication. Solve the Pretest 1 # 1–# 5
Lesson 2 Addition and Subtraction of decimals
Solve Review 1 A # 16 - # 20 Review 1 B # 16 - # 20
Lesson 3 Multiplication and power of decimals
Solve Review 1 A # 26 – 30 Solve Review 1 B # 26 – 30
Lesson 4 Division of decimals. Solve Review 1 A # 16 – 20,
Review 1 B # 16 - 20

Lesson 5 Table 3. Solve Mental Math # 5, Mental Math # 6
Solve Review 1 A # 6 – #10 Review # 6 - # 10
Lesson 6 Order of operations with decimals.
Solve Review 1 A # 16 – 20, Review 1 B # 16 – 20
Lesson 7 Laws for decimals
Solve the Review 1 A # 16 – # 20 Solve the Review 1 B # 16 – # 20
Lesson 8 Worksheets for Final Exam.
Solve Worksheet 1 A # 16 – # 20, Solve Worksheet 1 B # 16 – # 20

Lesson 9 Reviews
Review 3 A # 1 – # 10 all, Review 3 B # 1 – # 10 all
Lesson 10 Tests
Test 3 A # 1 – # 10 all, Test 1 B # 1 – # 10 all
Answers Chapter 3

Grade 7 Chapter 3 Lesson 1
Table 1
Multiplication and Division

$A + 0 = A$	$0 + A = A$	$A - 0 = A$	$A - A = 0$
$A \times 1 = A$	$1 \times A = A$	$A \div 1 = A$	$A \div A = 1$
$A \times 0 = 0$	$0 \times A = 0$	$0 \div A = 0$	$A \div 0 = \varnothing$

$A + B = B + A$ $A \times B = B \times A$ $\varnothing$ - no solutions

Number 5 from the left column multiply by number 4 from first row to get answer 20. $5 \times 4 = 20$ $20 \div 5 = 4$ $20 \div 4 = 5$

Memorize Table 1

×	1	2	3	4	5	6	7	8	9	10
1	1	2	3	4	5	6	7	8	9	10
2	2	4	6	8	10	12	14	16	18	20
3	3	6	9	12	15	18	21	24	27	30
4	4	8	12	16	20	24	28	32	36	40
5	5	10	15	20	25	30	35	40	45	50
6	6	12	18	24	30	36	42	48	54	60
7	7	14	21	28	35	42	49	56	63	70
8	8	16	24	32	40	48	56	64	72	80
9	9	18	27	36	45	54	63	72	81	90
10	10	20	30	40	50	60	70	80	90	100

Grade 7 Chapter 3 Lesson 1
Pretest
Evaluate

1) $22 \div 4.4 \times 5$

1) ______

2) $5.2 - 3.6 \div 4$

2) ______

3) $2^2 - 0.4^2 \div 0.2^2$

3) ______

4) $0.3^2 \div (0.6 - 0.3)^2$

4) ______

5) $4.8 \div (1.8^2 + 2.76)$

5) ______

Grade 7 Chapter 3 Lesson 2

Addition and Subtraction of Decimals

Decimals

A fraction is equal parts of whole number: $1 \div 4 = \frac{1}{4}$

A fraction contains a numerator (N) and a denominator (D).

The numerator written above the bar and shows how many equal parts are to be taken.

The denominator written under the bar and shows by how many equal parts the whole should be divided.

Fractions with a denominator of one and some zeros can represent like decimals: $\frac{3}{10} = 0.3$ $\frac{7}{100} = 0.07$ $1\frac{9}{1000} = 1.009$

Decimals contain two parts: to the left of the point stand digits of whole parts, to the right of the point stand digits of fractional parts, which called decimals digits.

Money can represent in decimals:
\$ 0.5 = \$ 0.50 = 50 c **\$ 0.05 = 5 c**

After the point, the 1 st place to the right is the tenth place, the 2nd place is the hundredth place, and 3rd is the thousandth place:

0.5 **3.28** **2.125**

Basic Property of Decimal

Adding or deleting zeros after the last digit to the right of the decimal point does not change the value of the decimal:

4 = 4. = 4.0 = 4.00 **5.00 = 5**

Grade 7 Chapter 3 Lesson 2

Rounding Decimals

Taking an approximate value for the number called rounding.
A rounded number is always rounded to a given place value.

A number can round to any digit: one, ten, hundred …

Rule for Rounding

If the digit to the right of the given place value is less than 5 (0, 1, 2, 3, 4) that digit and all digits to the right are replaced by zeros or dropped.

Example: rounding a number to the nearest ten: 283 $\approx$ 280 (3 < 5)

If the digit to the right of the given place value is greater than or equal to 5 (5, 6, 7, 8, 9) increase the digit in the given place by 1 and replace all other digits to the right by zeros or drop all digits to its right.

Example: rounding a number to the nearest ten: 238 $\approx$ 240 (8 > 5)

Comparison Decimals

To compare two decimals, make the number of decimals digits equal, add from the right the zeros and compare digits as whole and fractional part of decimals. (Work from left to right)

1.48 < 3.6 (1 < 3)
0.5 > 0.05

0.586 < 0.590 (8 < 9)
0.50 > 0.05

Addition and Subtraction of Decimals

To add or subtract decimals, line up the decimal points.
Then add or subtract from right to left.
Write a decimal point in the sum or difference.
Add zeroes as needed.

$$\begin{array}{r} 1.25 \\ +\ 0.80 \\ \hline 2.05 \end{array} \qquad \begin{array}{r} 1.25 \\ -\ 0.80 \\ \hline 0.45 \end{array}$$

Grade 7 Chapter 3 Lesson 2

Review 1 A

Rounding the decimals

to 0.1

1) 1.23
2) 2.34
3) 3.45
4) 4.56
5) 5.67

to 0.01

6) 1.234
7) 2.345
8) 3.456
9) 4.567
10) 5.678

Compare the decimals

Put symbols < , >, = instead question mark

11) 0.1 ? 0.01
12) 0.02 ? 0.2
13) 1.3 ? 1.03
14) 1.04 ? 1.4
15) 2.5 ? 2.05
16) 2.06 ? 2.6
17) 3.7 ? 3.07
18) 3.08 ? 3.8
19) 4.9 ? 4.09
20) 4.0 ? 4.9

Review 1 B

Add the decimals

1) 19.19 + 11.91
2) 28.28 + 12.82
3) 37.37 + 13.73
4) 46.46 + 14.64
5) 29.29 + 12.92
6) 38.38 + 13.83
7) 47.47 + 14.74
8) 56.56 + 15.65
9) 39.39 + 13.93
10) 48.48 + 14.84

Subtract the decimals

11) 31.1 – 19.19
12) 41.1 - 28.28
13) 51.1 - 37.37
14) 61.1 - 46.46
15) 42.21 - 29.29
16) 52.21 - 38.38
17) 62.21 - 47.47
18) 72.21 - 56.56
19) 53.32 - 39.39
20) 63.32 - 48.48

Grade 7 Chapter 3 Lesson 3
Multiplication and Power of decimals

Multiplication of Decimals

Multiply two decimals as if they are whole numbers.
The number of decimal digits in the product is the sum of the number of decimal digits in both factors:

1 digit	2 digits	3 digits
$\begin{array}{r} 2.5 \\ \times\ \ 4 \\ \hline 10.0 \end{array}$	$\begin{array}{r} 2.5 \\ \times 0.8 \\ \hline 2.00 \end{array}$	$\begin{array}{r} 1.25 \\ \times 0.8 \\ \hline 1.000 \end{array}$
$2.5 \times 4 = 10$	$2.5 \times 0.8 = 2$	$1.25 \times 0.8 = 1$

Multiplication of Decimal by Numbers 10, 100, 1000 …

To multiply a decimal by numbers 10, 100, or 1000 …move the decimal point to the right as many decimal places as there are zeros in the number:

1 zero – 1 place	2 zeros – 2 places	3 zeros – 3 places
$0.24 \times 10 = 2.4$	$0.24 \times 100 = 24$	$0.24 \times 1000 = 240$

Powers of Decimals

A power is multiplication of the same numbers:

$3.5^2 = 3.5 \times 3.5 = 12.25$ $\quad$ $0.5^3 = 0.5 \times 0.5 \times 0.5 = 0.25 \times 0.5 = 0.125$

$$\begin{array}{r} 3.5 \\ \times 3.5 \\ \hline 175 \\ +1050 \\ \hline 12.25 \end{array} \qquad \begin{array}{r} 0.5 \\ \times 0.5 \\ \hline 0.25 \end{array} \qquad \begin{array}{r} 0.25 \\ \times 0.5 \\ \hline 0.125 \end{array}$$

Grade 7 Chapter 3 Lesson 3

Review 1 A

Multiply the Decimals

1) 5×0.2	2) 0.5×2	3) 50×0.2
4) 500×0.2	5) 2.5×4	6) 2.5×0.4
7) 0.25×4	8) 0.25×0.4	9) 12.5×8
10) 1.25×8	11) 0.125×8	12) 2.5×8
13) 12.5×0.8	14) 1.25×0.8	15) 0.125×0.8
16) 0.25×8	17) 25×1.6	18) 35×1.6
19) 45×1.6	20) 2.5×1.6	21) 7.5×3.2
22) 7.5×6.4	23) 6.25×3.2	24) 6.25×0.32
25) 2.5×3.2	26) 2.5×0.2	27) 2.5×6.4
28) 2.5×0.64	29) 0.5×6.4	30) 0.5×64

Grade 7 Chapter 3 Lesson 3

Review 1 B

Find the powers of decimals

1) 0.1^2	2) 0.2^2	3) 0.3^2
4) 0.4^2	5) 0.5^2	6) 0.6^2
7) 0.7^2	8) 0.8^2	9) 0.9^2
10) 1.0^2	11) 1.1^2	12) 1.2^2
13) 1.3^2	14) 1.4^2	15) 1.5^2
16) 1.6^2	17) 1.7^2	18) 1.8^2
19) 1.9^2	20) 2.0^2	21) 2.1^2
22) 2.2^2	23) 2.3^2	24) 2.4^2
25) 2.5^2	26) 2.6^2	27) 2.7^2
28) 2.8^2	29) 2.9^2	30) 3.0^2

Grade 7 Chapter 3 Lesson 4

Division of Decimals

Division of a Decimal by a Whole Number

To divide a decimal by a natural number, first divide the whole part by the number, then in the quotient, place the point, after that, divide the fractional part of the number.

If the digit is not enough, add zeros to the right:

$2 \div 5 = 0.4$	$0.2 \div 5 = 0.04$	$0.02 \div 5 = 0.004$

Division of Decimal by Numbers 10, 100, and 1000…

To divide a decimal by numbers 10, 100, or 1000 …move the decimal point to the left as many decimal places as there are zeros in the number:

1 zero – 1 place	2 zeros – 2 places	3 zeros – 3 places
$12.5 \div 10 = 1.25$	$12.5 \div 100 = 0.125$	$12.5 \div 1000 = 0.0125$
$5 \div 10 = 0.5$	$5 \div 100 = 0.05$	$5 \div 1000 = 0.005$

Division of Decimals

To divide two decimals, multiply dividend and divisor by 10, 100, or 1000; the divisor has to be a whole number, after that, divide the decimal by the whole number:

$2 \div 0.5 = 20 \div 5 = 4$	$0.2 \div 0.5 = 2 \div 5 = 0.4$	$0.02 \div 0.5 = 0.2 \div 5 = 0.04$
$2 \times 10 = 20$	$0.2 \times 10 = 2$	$0.02 \times 10 = 0.2$
$0.5 \times 10 = 5$	$0.5 \times 10 = 5$	$0.5 \times 10 = 5$

Grade 7 Chapter 3 Lesson 4

$4 \div 0.8 = 40 \div 8 = 5$ $3.6 \div 0.9 = 36 \div 9 = 4$

Divide the numbers

Review 1 A

1) $1 \div 5$	2) $2 \div 5$	3) $3 \div 5$	4) $4 \div 5$
5) $6 \div 5$	6) $7 \div 5$	7) $8 \div 5$	8) $9 \div 5$
9) $11 \div 2$	10) $13 \div 2$	11) $17 \div 2$	12) $19 \div 2$
13) $11 \div 5$	14) $23 \div 5$	15) $37 \div 5$	16) $49 \div 5$
17) $1 \div 4$	18) $2 \div 4$	19) $3 \div 4$	20) $5 \div 4$

Review 1 B

1) $1 \div 0.8$	2) $3 \div 0.8$	3) $5 \div 0.8$	4) $7 \div 0.8$
5) $24 \div 1.6$	6) $40 \div 1.6$	7) $80 \div 1.6$	8) $96 \div 1.6$
9) $32 \div 1.6$	10) $48 \div 1.6$	11) $56 \div 1.6$	12) $64 \div 1.6$
13) $5.6 \div 1.6$	14) $7.2 \div 1.6$	15) $8.8 \div 1.6$	16) $10.4 \div 1.6$
17) $12.8 \div 3.2$	18) $16 \div 3.2$	19) $19.2 \div 3.2$	20) $25.6 \div 3.2$

Grade 7 Chapter 3 Lesson 5
Table 3 Decimals

Properties of Decimal

$4 = 4. = 4.0 = 4.00$ $\qquad$ $5.00 = 5$

Rounding of Decimals

$2.83 \approx 2.8$ $\qquad$ $2.38 \approx 2.4$

Comparison of Decimals

$1.48 < 3.6$ $\qquad$ $0.5 > 0.05$

Addition and Subtraction of Decimals

$$\begin{array}{r} 1.25 \\ +\ 0.80 \\ \hline 2.05 \end{array} \qquad \begin{array}{r} 1.25 \\ -\ 0.80 \\ \hline 0.45 \end{array}$$

Multiplication of Decimals

$0.24 \times 10 = 2.4$ $\qquad$ $0.24 \times 100 = 24$ $\qquad$ $0.24 \times 1000 = 240$

$2.5 \times 4 = 10$ $\qquad$ $1.25 \times 8 = 10$ $\qquad$ $0.5 \times 0.2 = 0.1$

Powers of Decimals

$3.5^2 = 3.5 \times 3.5 = 12.25$ $\qquad$ $0.5^3 = 0.5 \times 0.5 \times 0.5 = 0.25 \times 0.5 = 0.125$

Division of Decimals

$5 \div 10 = 0.5$ $\qquad$ $5 \div 100 = 0.05$ $\qquad$ $5 \div 1000 = 0.005$

$2 \div 5 = 0.4$ $\qquad$ $0.2 \div 5 = 0.04$ $\qquad$ $0.02 \div 5 = 0.004$

$2 \div 0.5 = 20 \div 5 = 4$ $\qquad$ $0.2 \div 0.5 = 2 \div 5 = 0.4$ $\qquad$ $0.02 \div 0.5 = 0.2 \div 5 = 0.04$

Order of Operations with the Same Level

$0.8 + 0.6 + 0.2 = 0.8 + 0.2 + 0.6 = 1.6$ $\qquad$ $2.5 \times 6 \times 4 = 2.5 \times 4 \times 6 = 60$

$0.8 - 0.2 + 0.4 = 0.6 + 0.4 = 1$ $\qquad$ $1.6 \div 4 \times 2 = 0.4 \times 2 = 0.8$

Order of Operations of Two Levels

$0.2 \times 4 - 0.2 = 0.8 - 0.2 = 0.6$ $\qquad$ $1.8 \div 9 + 0.4 = 0.2 + 0.4 = 0.6$

$0.6 + 0.4 \times 2 = 0.6 + 0.8 = 1.4$ $\qquad$ $0.8 - 1.5 \div 3 = 0.8 - 0.5 = 0.3$

Order of Operations in Expression with Parentheses

$(0.5 + 0.4) \times 2 = 0.9 \times 2 = 1.8$ $\qquad$ $3 \times (0.9 - 0.7) = 3 \times 0.2 = 0.6$

$(1.8 - 0.4) \div 7 = 1.4 \div 7 = 0.2$ $\qquad$ $2 \div (0.4 - 0.2) = 2 \div 0.2 = 10$

Groping Symbols: 1) Parentheses ( ) 2) Bracket [ ] 3) Brace { }

Order of Five Operations

Grouping Symbols (1) [2] {3} Exponents (Power) a^n

Multiplication or Division from left to right $\times$ or $\div$ L → R

Addition or Subtraction from left to right + or – L → R

Grade 7 Chapter 3 Lesson 5

Mental Math 5

Decimals

1) $0.2 + 0$
2) $0 + 1.5$
3) $0.2 + 0.5$
4) $0.5 + 0.2$
5) $1.8 + 0.2$
6) $0.6 - 0$
7) $1.7 - 1.7$
8) $0.5 - 0.2$
9) 0.6×0
10) 0×1.7
11) 1×1.9
12) 0.8×1
13) 0.2×5
14) 0.4×5
15) 5×0.6
16) 0.5×8
17) 1.2×5
18) 5×1.4
19) $0 \div 0.6$
20) $0.7 \div 0$
21) $0.8 \div 1$
22) $1.9 \div1.9$
23) 0.2^2
24) 0.3^2
25) 0.4^2
26) 0.5^2
27) 0.6^2
28) 0.7^2
29) 0.8^2
30) 0.9^2

Mental Math 6

Laws for Whole Numbers

1) $8\times3+2\times3$
2) $7\times5-2\times5$
3) $7\times4+3\times4$
4) $6\times2-4\times2$
5) $4\times5+6\times5$
6) $9\times4-8\times4$
7) $5\times2+5\times8$
8) $3\times7-3\times6$
9) $3\times4+3\times6$
10) $5\times9-5\times8$
11) $8\times75+8\times25$
12) $25\times6-15\times6$
13) $2\times17+2\times13$
14) $75\times4-35\times4$
15) $3\times27+3\times23$
16) $93\times 2-2\times53$
17) $14\times4+4\times16$
18) $6\times28-6\times18$
19) $29\times3+21\times3$
20) $67\times2-2\times47$

Grade 7 Chapter 3 Lesson 5

Laws for Decimals
Review 1 A

1) $80 \times 7.5 + 80\times 2.5$
2) $3.7 \times 40 + 40 \times 1.3$

3) $8.5 \times 40 - 4.5 \times 40$
4) $5.4 \times 50 - 3.4 \times 50$

5) $70\times 3.7 + 6.3 \times 70$
6) $4.9 \times 30 + 30 \times 5.1$

7) $7.5 \times 60 - 3.5 \times 60$
8) $6.9 \times 80 - 80 \times 1.9$

9) $30 \times 2.7 + 30 \times 2.3$
10) $2.4 \times 30 + 1.6 \times 30$

Review 1 B

11) $9.3 \times 20 - 20 \times 5.3$
12) $5.8 \times 50 + 4.2 \times 50$

13) $4.3 \times 40 + 40 \times 2.7$
14) $9.6 \times 20 - 2.6 \times 20$

15) $50 \times 6.8 - 50 \times 1.8$
16) $7.8 \times 40 - 3.8 \times 40$

17) $2.9 \times 30 + 3.1 \times 30$
18) $3.3 \times 60 + 2.7 \times 60$

19) $9.7 \times 50 - 50 \times 2.7$
20) $80 \times 9.7 - 80 \times 5.7$

Grade 7 Chapter 3 Lesson 6

Order of Operations with Decimals

If an expression has five operations, first work on operations at the third level (a^n), after that, work on operations at the second level (× or÷), last, work on operations at the first level (+ or -) in order, as they occur, from left to right.

in short

Grouping Symbols (1) [2] {3}

Exponents or Power – third level of operations a^n

Multiplication or Division from left to right–second level of operations

× or ÷ L → R

Addition or Subtraction from left to right – first level of operations

\+ or – L → R

Examples:

a) $0.2^3 \div (0.4 - 0.2)^2 = 0.008 \div (0.2)^2 = 0.008 \div 0.04 = 0.8 \div 4 = 0.2$

b) $2^2 - 0.8^2 \div 0.4^2 = 4 - 0.64 \div 0.16 = 4 - 4 = 0$

c) $3^2 - 0.3^4 \div 0.3^3 = 9 - 0.0081 \div 0.027 = 9 - 8.1 \div 27 = 9 - 0.3 = 8.7$

d) $0.8 \times 2 + (1.5^2 - 0.25) = 1.6 + (2.25 - 0.25) = 1.6 + 2 = 3.6$

e) $5\times [16.72 \div 8 - (40 - 39.01)] = 5\times [2.09 - 0.99] = 5 \times 1.1 = 5.5$

f) $3.7\div [6 - (0.3+0.2\times 10)] = 3.7\div [6 - (0.3+2)] = 3.7\div [6-2.3] = 3.7\div 3.7=1$

g) $30\div [5.5 - (1.5^2 + 1.75)] = 30\div [5.5 - (2.25 + 1.75)] = 30\div [5.5 - 4] = 20$

h) $9.1\div \{1 - [0.3(0.2 + 0.1)]\} = 9.1\div \{1 - [0.3\times 0.3]\} = 9.1\div \{1 - 0.09\} = 10$

i) $2 \times 0.3 + \{4.9 \div [1.2 - (0.2 + 0.3)]\} = 0.6 + \{4.9 \div [1.2 - 0.5]\} =$
$0.6 + \{4.9 \div 0.7\} = 0.6 + 7 = 7.6$

j) $0.5\ [1.2 - (0.6 + 0.3)] - 0.2^2 = 0.5\ [1.2 - 0.9] - 0.04 =$
$0.5 \times 0.3 - 0.04 = 0.15 - 0.04 = 0.11$

Grade 7 Chapter 3 Lesson 6

Review 1 A

Perform the operations

1) $3 - 0.3 \div 0.3$
2) $1.1 \times 6 - 6.6 \div 2$

3) $9.9 \div 3 - 1.1 \times 3$
4) $2.8 \div 4 - 0.2 \times 2$

5) $1.8 \div 0.6 + 0.3 \times 2$
6) $2.5 \div 0.5 + 2.5 \times 6$

7) $(4.4 - 1.4) \times 2 \div 3$
8) $0.1 \div (10 - 3 \times 3)$

9) $5(0.2)(5)(0.4)$
10) $(0.5 + 0.4) \cdot 2 \div 3$

11) $2^2 - 0.8^2 \div 0.4^2$
12) $3^2 - 0.3^2 \div 0.3$

13) $5 \times 0.8 \times 2.5 \times 0.4$
14) $0.5 \times 6 \times 0.4 \times 2.5$

15) $4 \times 4 \times 0.5 \times 2.5$
16) $3 \times (0.9 - 0.7) \div 0.6$

17) $(0.8 - 0.3) \times 2 \div 5$
18) $0.3 \div (1.5 + 1.5) \times 2$

19) $2 \div (0.4 - 0.2) \times 0.1$
20) $(0.6 + 0.4) \div 10 \times 5$

Review 1 B

1) $(1.8 - 0.4) \div 7 \times 5$
2) $4 \times (2.5 - 1.5 \div 3)$

3) $4 - 2.6 \div (4.2 - 2.2)$
4) $(5.1 \times 4 - 19.4) \div 0.5$

5) $24 \div (3.5 + 6.5) \div 1.2$
6) $2.4 \div (6.4 - 4.4) \times 5$

7) $1.6 \div (4.6 + 3.4) \div 2$
8) $6.9 \div (0.3 + 0.2 \times 10)$

9) $5(1.6 \div 8 - 0.1)$
10) $(2 \div 0.4 - 7 \div 3.5) \div 1.5$

11) $4.9 \div [1.2 - (0.2 + 0.3)]$
12) $(1.4^2 - 1.2^2) \times 5$

13) $(1.7^2 - 1.5^2) \times 5$
14) $(1.1^2 + 1.3^2) + 1.1$

15) $(1.8^2 + 1.6^2) \times 10$
16) $0.2^3 \div (0.4 - 0.2)^2$

17) $5(1.2 - 0.9) \times 10$
18) $2 \div (1.5^2 + 1.75)$

19) $1.6 \div (1.5^2 - 0.25)$
20) $(2 \div 0.4 - 7 \div 3.5) \div 1.5$

Grade 7 Chapter 3 Lesson 7

Laws for Decimals

Addition

$A + 0 = A$ $0.2 + 0 = 0.2$ $0 + A = A$ $0 + 1.3 = 1.3$

$A + B = B + A$ $0.2 + 0.3 = 5$ $0.3 + 0.2 = 0.5$ $0.2 + 0.3 = 0.3 + 0.2$

$A + (B + C) = (A + B) + C$ $1.7 + 0.2 = 1.0 + 0.7 + 0.2 = 1.0 + 0.9 = 1.9$

Subtraction

$A – 0 = A$ $0.6 – 0 = 0.6$ $A – A = 0$ $1.7 – 1.7 = 0$

$0.2 + 0.3 = 0.5$ $0.5 – 0.2 = 0.3$ $0.5 – 0.3 = 0.2$

Multiplication

$A \times 0 = 0$ $0.6 \times 0 = 0$ $0 \times A = 0$ $0 \times 1.7 = 0$

$A \times 1 = A$ $0.8 \times 1 = 0.8$ $1 \times A = A$ $1 \times 1.9 = 1.9$

$A B = B A$ $0.2 \times 0.3 = 0.06$ $0.3 \times 0.2 = 0.06$ $0.2 \times 0.3 = 0.3 \times 0.2$

$A (BC) = (AB)C$ $0.5 (2 \times 4) = (0.5 \times 2) \times 4 = 1 \times 4 = 4$

$(B + C) A = AB + AC$ $(1 + 0.4)5 = 1 \times 5 + 0.4 \times 5 = 5 + 2 = 7$

$A (B - C) = AB – AC$ $5(1 – 0.4) = 5 \times 1 – 5 \times 0.4 = 5 – 2 = 3$

Grade 7 Chapter 3 Lesson 7

Division

$0 \div A = 0$ $\quad 0 \div 0.6 = 0$ $\quad A \div 0$ None $\quad 0.7 \div 0$ None

$A \div 1 = A$ $\quad 0.8 \div 1 = 0.8$ $\quad A \div A = 1$ $\quad 1.9 \div 1.9 = 1$

$(A + B) \div C = A \div C + B \div C$

$(1.8 + 1.2) \div 3 = 1.8 \div 3 + 1.2 \div 3 = 0.6 + 0.4 = 1$

$(A - B) \div C = A \div C - B \div C$

$(1.8 - 1.2) \div 3 = 1.8 \div 3 - 1.2 \div 3 = 0.6 - 0.4 = 0.2$

$(AB) \div C = (A \div C)\,B$ $\quad (0.6 \times 7) \div 0.3 = (0.6 \div 0.3) \times 7 = 2 \times 7 = 14$

$(AB) \div C = A(B \div C)$ $\quad (7 \times 0.6) \div 0.3 = 7\,(0.6 \div 0.3) = 7 \times 2 = 14$

Mental Calculations

$2.4 \times 5 = 2.4 \div 2 \times 10 = 1.2 \times 10 = 12$

$2.4 \times 25 = 2.4 \div 4 \times 100 = 0.6 \times 100 = 60$

$2.4 \times 125 = 2.4 \div 8 \times 1000 = 0.3 \times 1000 = 300$

Common Factor

$(B + C)\,A = AB + AC$ $\quad (0.4 + 0.1)5 = 0.4 \times 5 + 0.1 \times 5 = 2 + 0.5 = 2.5$

$AB + AC = A(B + C)$ $\quad 0.4 \times 5 + 0.1 \times 5 = 5\,(0.4 + 0.1) = 5(0.5) = 2.5$

$A(B - C) = AB - AC$ $\quad 5\,(0.4 - 0.1) = 5 \times 0.4 - 5 \times 0.1 = 2 - 0.5 = 1.5$

$AB - AC = A(B - C)$ $\quad 5 \times 0.4 - 5 \times 0.1 = 5\,(0.4 - 0.1) = 5(0.3) = 1.5$

Grade 7 Chapter 3 Lesson 7

Review 1 A

$$5 \times 0.6 + 5 \times 0.4 = 5\,(0.6 + 0.4) = 5(1) = 5$$

Simplify by laws for decimals

1) $8 \times 7.5 + 8 \times 2.5$

2) $2.5 \times 6 - 1.5 \times 6$

3) $3 \times 3.7 + 6.3 \times 3$

4) $7.5 \times 4 - 3.5 \times 4$

5) $3 \times 2.7 + 3 \times 2.3$

6) $9.3 \times 2 - 2 \times 5.3$

7) $1.4 \times 4 + 4 \times 1.6$

8) $6 \times 2.8 - 6 \times 1.8$

9) $2.9 \times 3 + 2.1 \times 3$

10) $7 \times 2.4 - 1.4 \times 7$

11) $3.7 \times 4 + 4 \times 1.3$

12) $5.4 \times 5 - 3.4 \times 5$

13) $4.9 \times 2 + 2 \times 5.1$

14) $3.9 \times 6 - 6 \times 1.9$

15) $2.4 \times 9 + 1.6 \times 9$

16) $5.8 \times 3 + 3.2 \times 3$

17) $9.6 \times 2 - 2.6 \times 2$

18) $3.8 \times 3 - 1.8 \times 3$

19) $0.3 \times 6 + 0.7 \times 6$

20) $7 \times 5.7 - 7 \times 4.7$

Grade 7 Chapter 3 Lesson 7

Review 1 B

$8 \times 2.5 - 8 \times 1.5 = 8\,(2.5 - 1.5) = 8(1) = 8$

Simplify by laws for decimals

1) $2 \times 0.7 + 2 \times 0.3$

2) $4 \times 6.4 - 4 \times 5.4$

3) $0.3 \times 7.8 + 0.3 \times 2.2$

4) $4 \times 3.2 - 4 \times 2.2$

5) $3.6 \times 2.3 + 3.6 \times 7.7$

6) $5 \times 3.6 + 5 \times 6.4$

7) $7.9 \times 4.7 + 7.9 \times 5.3$

8) $3 \times 5.6 + 4.4 \times 3$

9) $3.7 \times 2.4 + 3.7 \times 7.6$

10) $2 \times 6.2 - 2 \times 5.2$

11) $5.8 \times 7.3 + 5.8 \times 2.7$

12) $7 \times 4.8 - 7 \times 3.8$

13) $6.1 \times 3.4 + 6.1 \times 6.6$

14) $9 \times 3.8 - 9 \times 2.8$

15) $6.7 \times 8.4 + 6.7 \times 1.6$

16) $3 \times 4.1 - 3 \times 3.1$

17) $4.3 \times 3 + 3.5 \times 3 - 5.8 \times 3$

18) $4.9 \times 7 - 1.8 \times 7 + 6.9 \times 7$

19) $5.6 \times 4 + 2.7 \times 4 - 8.3 \times 4$

20) $3.4 \times 5 - 1.7 \times 5 + 8.3 \times 5$

Grade 7 Chapter 3 Lesson 8

Worksheet 1 A for Final Exam

Whole Numbers

Perform the indicated operations

1) $(4^2 - 2^2) \div 6$

2) $(10^2 - 9^2) \div 19$

3) $(7^2 - 5^2) \div 8$

4) $(11^2 - 9^2) \div 20$

5) $(8^2 - 5^2) \div 3$

6) $(13^2 - 9^2) \div 8$

7) $84 \div (8^2 - 6^2)$

8) $81 \div (14^2 - 13^2)$

9) $52 \div (7^2 - 6^2)$

10) $64 \div (9^2 - 7^2)$

11) $96 \div (8^2 - 4^2)$

12) $56 \div (4^2 - 3^2)$

13) $(9^2 - 6^2) \div (8^2 - 7^2)$

14) $(5^2 - 4^2)(3^2 - 2^2)$

15) $2^2(10 - 2^2 \div 2)$

16) $2^3 \div 2^2 \times 2 + 2^2 \times 5$

17) $5(3^2 + 2^3 \div 2^2)$

18) $(2 + 4)^2 \div 3 \times 2$

19) $2[2^2 + (1 + 2)^2]$

20) $2^2[2^2 + (2 + 2)^2]$

Grade 7 Chapter 3 Lesson 8

Worksheet 1 B for Final Exam

Decimals

Perform the indicated operations

1) $0.2^2 \div 0.2^2 - 0.2$

2) $0.3^3 \div 0.3^2 - 0.3$

3) $0.4^2 \div 0.2^2 - 2$

4) $0.4^2 \div 0.2^2 - 4$

5) $0.6^2 \div 0.2^2 - 4$

6) $2^2 - 0.8^2 \div 0.4^2$

7) $2^2 - 0.4^2 \div 0.2^2$

8) $5^2 - 0.6^2 \div 0.2^2$

9) $3^2 - 0.8^2 \div 0.4^2$

10) $7^2 - 0.9^2 \div 0.3^2$

11) $0.2^2 \div (0.4 - 0.3)^2$

12) $0.3^3 \div (0.6 - 0.3)^2$

13) $0.4^2 \div (0.6 - 0.4)^2$

14) $0.6^2 \div (0.5 - 0.2)^2$

15) $0.8^2 \div (0.7 - 0.3)^2$

16) $(0.4 - 0.2)^2 \div 0.2^2$

17) $(0.6 - 0.3)^2 \div 0.3$

18) $(0.8 - 0.4)^2 \div 0.4$

19) $(0.5 - 0.2)^2 \div 0.9$

20) $(0.7 - 0.5)^2 \div 0.4$

Grade 7 Chapter 3 Lesson 9
Reviews for Chapter 3
Review 3 A

Read directions carefully. Answers must be in the indicated place.
Show all steps for full credits. Do only 1 operation per 1 step.
Solve 1 problem 5 minutes. Each problem is worth 5 points.

Evaluate

1) **Add decimals** $37.37 + 13.73$

1) ______

2) **Subtract decimals** $61.1 - 46.46$

2) _______

3) **Multiply decimals** 0.75×4.8

3) ______

4) **Divide decimals** $36.45 \div 0.9$

4) ______

5) **Evaluate** $9.5\times5.8 - 9.5\times4.8$

5)______

Grade 7 Chapter 3 Lesson 9
Review 3 A

Read directions carefully. Answers must be in the indicated place.
Show all steps for full credits. Do only 1 operation per 1 step.
Solve 1 problem 5 minutes. Each problem is worth 5 points.

Evaluate

6) $33 \div 6.6 \times 5$

6) ______

7) $9.6 - 6.8 \div 4$

7) ______

8) $6^2 - 0.6^2 \div 0.3^2$

8) ______

9) $0.4^2 \div (0.6 - 0.4)^2$

9) ______

10) $12 \div (2.6^2 + 1.24)$

10) ______

Grade 7 Chapter 3 Lesson 9
Review 3 B

Read directions carefully. Answers must be in the indicated place.
Show all steps for full credits. Do only 1 operation per 1 step.
Solve 1 problem 5 minutes. Each problem is worth 5 points.

Evaluate

1) Add decimals $46.46 + 14.64$

1)______

2) Subtract decimals $51.1 - 37.37$

2)______

3) Multiply decimals 0.75×2.4

3) ______

4) Divide decimals $45.54 \div 0.9$

4) _____

5) Evaluate $8.6\times4.7 - 8.6\times3.7$

5) ______

Grade 7 Chapter 3 Lesson 9
Review 3 B

Read directions carefully. Answers must be in the indicated place.
Show all steps for full credits. Do only 1 operation per 1 step.
Solve 1 problem 5 minutes. Each problem is worth 5 points.

Evaluate

6) $33 \div 5.5 \times 6$

6) ______

7) $4.8 - 3.4 \div 2$

7) ______

8) $4^2 - 0.4^2 \div 0.2^2$

8) ______

9) $0.6^2 \div (0.5 - 0.3)^2$

9) ______

10) $9.9 \div (2.8^2 + 1.16)$

10) ______

Grade 7 Chapter 3 Lesson 10
Tests for Chapter 3
Test 3 A

Read directions carefully. **Answers must be in the indicated place.**
Show all steps for full credits. **Do only 1 operation per 1 step.**
Solve 1 problem 5 minutes. **Each problem is worth 5 points.**

Evaluate

1) **Add decimals** $19.19 + 11.91$

1) ______

2) **Subtract decimals** $41.1 - 28.28$

2) ______

3) **Multiply decimals** 0.25×1.6

3) ______

4) **Divide decimals** $18.3 \div 0.6$

4) ______

5) **Evaluate** $7.8\times6.2 - 7.8\times5.2$

5) ______

Grade 7 Chapter 3 Lesson 10
Test 3 A

Read directions carefully. **Answers must be in the indicated place.**
Show all steps for full credits. **Do only 1 operation per 1 step.**
Solve 1 problem 5 minutes. **Each problem is worth 5 points.**

Evaluate

6) $22 \div 5.5 \times 4$

6) ______

7) $4.8 - 2.4 \div 2$

7) ______

8) $3^2 - 0.9^2 \div 0.3^2$

8) ______

9) $0.2^2 \div (0.4 - 0.2)^2$

9) ______

10) $4.4 \div (1.6^2 + 1.44)$

10)______

Grade 7 Chapter 3 Lesson 10

Test 3 B

Read directions carefully. Answers must be in the indicated place.
Show all steps for full credits. Do only 1 operation per 1 step.
Solve 1 problem 5 minutes. Each problem is worth 5 points.

Evaluate

1) **Add decimals** $28.28 + 12.82$

1) ______

2) **Subtract decimals** $31.1 - 19.19$

2) ______

3) **Multiply decimals** 0.25×3.2

3) ______

4) **Divide decimals** $24.3 \div 0.6$

4) ______

5) **Evaluate** $6.8\times3.9 - 6.8\times2.9$

5) _____

Grade 7 Chapter 3 Lesson 10
Test 3 B

Read directions carefully. Answers must be in the indicated place.
Show all steps for full credits. Do only 1 operation per 1 step.
Solve 1 problem 5 minutes. Each problem is worth 5 points.
Evaluate

6) $22 \div 4.4 \times 5$

6) ______

7) $5.2 - 3.6 \div 4$

7) ______

8) $2^2 - 0.4^2 \div 0.2^2$

8) ______

9) $0.3^2 \div (0.6 - 0.3)^2$

9) ______

10) $4.8 \div (1.8^2 + 2.76)$

10) _____

Grade 7 Answers Chapter 3

Lesson 1

1)	25	2)	4.3	3)	0	4)	1	5)	0.8

Lesson 2 Review 1 A

1)	1.2	2)	2.3	3)	3.5	4)	4.6	5)	5.7
6)	1.23	7)	2.35	8)	3.46	9)	4.57	10)	5.68
11)	>	12)	<	13)	>	14)	<	15)	>
16)	<	17)	>	18)	<	19)	>	20)	<

Review 1 B

1)	31.1	2)	41.1	3)	51.1	4)	61.1	5)	42.21
6)	52.21	7)	62.21	8)	72.21	9)	53.32	10)	63.32
11)	11.91	12)	12.82	13)	13.73	14)	14.64	15)	12.92
16)	13.83	17)	14.74	18)	15.65	19)	13.93	20)	14.84

Lesson 3 Review 1 A

1)	1	2)	1	3)	10	4)	100	5)	10
6)	1	7)	1	8)	0.1	9)	100	10)	10
11)	1	12)	20	13)	10	14)	1	15)	0.1
16)	2	17)	40	18)	56	19)	72	20)	4
21)	24	22)	48	23)	20	24)	2	25)	8
26)	0.5	27)	16	28)	1.6	29)	3.2	30)	32

Lesson 3 Review 1 B

1)	0.01	2)	0.04	3)	0.09	4)	0.16	5)	0.25
6)	0.36	7)	0.49	8)	0.64	9)	0.81	10)	1
11)	1.21	12)	1.44	13)	1.69	14)	1.96	15)	2.25
16)	2.56	17)	2.89	18)	3.24	19)	3.61	20)	4
21)	4.41	22)	4.84	23)	5.29	24)	5.76	25)	6.25
26)	6.76	27)	7.29	28)	7.84	29)	8.41	30)	9

Grade 7 Answers Chapter 3

Lesson 4 Review 1 A

1) 0.2	2) 0.4	3) 0.6	4) 0.8	5) 1.2
6) 1.4	7) 1.6	8) 1.8	9) 5.5	10) 6.5
11) 8.5	12) 9.5	13) 2.2	14) 4.6	15) 7.4
16) 9.8	17) 0.25	18) 0.5	19) 0.75	20) 1.25

Review 4 Review 1 B

1) 1.25	2) 3.75	3) 6.25	4) 8.75	5) 15
6) 25	7) 50	8) 60	9) 20	10) 30
11) 35	12) 40	13) 3.5	14) 4.5	15) 5.5
16) 6.5	17) 4	18) 5	19) 6	20) 8

Lesson 5 Mental Math 5

1) 0.2	2) 1.5	3) 0.7	4) 0.7	5) 2
6) 0.6	7) 0	8) 0.3	9) 0	10) 0
11) 1.9	12) 0.8	13) 1	14) 2	15) 3
16) 4	17) 6	18) 7	19) 0	20) Ø
21) 0.8	22) 1	23) 0.04	24) 0.09	25) 0.16
26) 0.25	27) 0.36	28) 0.49	29) 0.64	30) 0.81

Lesson 5 Mental Math 6

1) 30	2) 25	3) 40	4) 4	5) 50
6) 4	7) 50	8) 3	9) 30	10) 5
11) 800	12) 60	13) 60	14) 160	15) 150
16) 80	17) 120	18) 60	19) 150	20) 40

Lesson 5 Laws for decimals

Review 1 A

1) 800	2) 200	3) 160	4) 100	5) 700
6) 300	7) 240	8) 400	9) 150	10) 120

Review 1 B

1) 80	2) 500	3) 280	4) 140	5) 250
6) 160	7) 180	8) 360	9) 350	10) 320

Grade 7 Answers Chapter 3

Lesson 6 Review 1 A

1)	2	2)	3.3	3)	0	4)	0.3	5)	3.6
6)	20	7)	2	8)	0.1	9)	2	10)	0.6
11)	0	12)	8.7	13)	4	14)	3	15)	20
16)	1	17)	0.2	18)	0.2	19)	1	20)	0.5

Lesson 6 Review 1 B

1)	1	2)	8	3)	2.7	4)	2	5)	2
6)	6	7)	0.1	8)	3	9)	0.5	10)	2
11)	7	12)	2.6	13)	3.2	14)	4	15)	58
16)	0.2	17)	15	18)	0.5	19)	0.8	20)	2

Lesson 7 Review 1 A

1)	80	2)	6	3)	30	4)	16	5)	15
6)	8	7)	12	8)	6	9)	15	10)	7
11)	20	12)	10	13)	20	14)	12	15)	36
16)	27	17)	14	18)	6	19)	6	20)	7

Lesson 7 Review 1 B

1)	2	2)	4	3)	3	4)	4	5)	36
6)	50	7)	79	8)	30	9)	37	10)	2
11)	58	12)	7	13)	61	14)	9	15)	67
16)	3	17)	6	18)	70	19)	0	20)	50

Lesson 8 Worksheet 1 A

1)	2	2)	1	3)	3	4)	2	5)	13
6)	11	7)	3	8)	3	9)	4	10)	2
11)	2	12)	8	13)	3	14)	45	15)	32
16)	24	17)	55	18)	24	19)	26	20)	80

Grade 7 Answers Chapter 3

Lesson 8 Worksheet 1 B

1)	0.8	2)	0	3)	2	4)	0	5)	5
6)	0	7)	0	8)	16	9)	5	10)	40
11)	4	12)	0.3	13)	4	14)	4	15)	4
16)	1	17)	0.3	18)	0.4	19)	0.1	20)	0.1

Lesson 9 Review 3 A

1)	51.1	2)	14.64	3)	3.6	4)	40.5	5)	9.5
6)	25	7)	7.9	8)	32	9)	4	10)	1.5

Lesson 9 Review 3 B

1)	61.1	2)	13.73	3)	1.8	4)	50.6	5)	8.6
6)	36	7)	3.1	8)	12	9)	9	10)	1.1

Lesson 10 Test 3 A

1)	31.1	2)	12.82	3)	0.4	4)	30.5	5)	7.8
6)	16	7)	3.6	8)	0	9)	1	10)	1.1

Lesson 10 Test 3 B

1)	41.1	2)	11.91	3)	0.8	4)	40.5	5)	6.8
6)	25	7)	4.3	8)	0	9)	1	10)	0.8

Grade 7

Chapter 4

Fractions and Mixed Numbers

Grade 7 Chapter 4 Contents
Fractions and Mixed Numbers
Lesson 1 – 10
Basic Problems

Work with lesson 1, read rules and examples, 2) Solve the problems from contents, 3) Check answers in the end of chapter 1, 4) Work with the next lessons 2, 3, 4, 5, …, 5) Solve Test 1 A, 6) Grading Scale: A =10 points, B=9points, C=8 points, D=7 points, F=0-6 points, 7)If you solve 10 problems correctly, you get an A, 8) If have an A, you can move on to next chapter, 9)If your grade is lower than an A, work with the mistakes, then solve Test 1B, 10) When you finish all of the chapters, solve Final Exam in last chapter.

Grade 7 Chapter 4 Lesson 1
Table 1
Multiplication and Division

A + 0 = A **0 + A = A** **A – 0 = A** **A – A = 0**
A × 1 = A **1 × A = A** **A ÷ 1 = A** **A ÷ A = 1**
A × 0 = 0 **0 × A = 0** **0 ÷ A = 0** **A ÷ 0 = Ø**
A + B = B + A **A × B = B × A** **Ø no solution**

Number 5 from the left column multiply by number 4 from first row to get answer 20.
5 × 4 = 20 20 ÷ 5 = 4 20 ÷ 4 = 5

Memorize Table 1

×	1	2	3	4	5	6	7	8	9	10
1	1	2	3	4	5	6	7	8	9	10
2	2	4	6	8	10	12	14	16	18	20
3	3	6	9	12	15	18	21	24	27	30
4	4	8	12	16	20	24	28	32	36	40
5	5	10	15	20	25	30	35	40	45	50
6	6	12	18	24	30	36	42	48	54	60
7	7	14	21	28	35	42	49	56	63	70
8	8	16	24	32	40	48	56	64	72	80
9	9	18	27	36	45	54	63	72	81	90
10	10	20	30	40	50	60	70	80	90	100

Grade 7 Chapter 4 Lesson 1
Pretest Chapter 4

1) Divide numbers

$$\frac{3}{8} \div \frac{3}{4} \div 1\frac{1}{4}$$

1) ______

2) Evaluate

$$(1\frac{4}{5}) \div \left(1\frac{1}{2}\right)^2$$

2) ______

3) Evaluate

$$\left(\frac{1}{5} - \frac{1}{10}\right)^2 \div \frac{1}{10}$$

3) ______

4) Evaluate

$$1\frac{1}{2} \div (5\ \frac{1}{2} 4\ \frac{1}{4})$$

4) _____

5) Evaluate

$$1\frac{3}{4} \times 12 - 1\frac{1}{4} \times 12$$

5) ______

Grade 7 Chapter 4 Lesson 2

Fractions and Mixed Numbers

A fraction is equal parts of whole number: $1 \div 4 = \frac{1}{4}$

A fraction contains a numerator (N) and a denominator (D).

The numerator written above the bar and shows how many equal parts are to be taken.

The denominator written under the bar and shows by how many equal parts the whole should be divided.

A fraction is a quotient of division, where the numerator of the fraction is the dividend, and the denominator of the fraction is the divisor.

The bar of the fraction is another sign of division.
Fractional expressions written with a bar:

$3 \div 4 = \frac{3}{4}$ $N \div D = \frac{N}{D}$

The fraction called proper if the numerator of the fraction is less than the denominator of the fraction.
A proper fraction is always less than one:

$N < D$ $\frac{N}{D} < 1$ $3 < 4$ $\frac{3}{4} < 1$

The fraction called improper if the numerator of the fraction equal to or more than the denominator of the fraction.
If the numerator is equal to the denominator, then the fraction equals one. $N = D$ $\frac{4}{4} = 1$

Grade 7 Chapter 4 Lesson 2

If the numerator is greater than the denominator, then the fraction is greater than one:

N > D $\frac{5}{4} > 1$ $5 > 4$

A mixed number contains a whole number and a fraction together:

$2 + \frac{1}{2} = 2\frac{1}{2}$ $\frac{1}{4} + 1 = 1\frac{1}{4}$

To change the mixed number to the improper fraction, multiply the denominator by the whole number and add the numerator. Put this result in as the new numerator and the denominator remains the same: $1\frac{3}{4} = \frac{4 \cdot 1 + 3}{4} = \frac{7}{4}$ $2\frac{1}{3} = \frac{3 \cdot 2 + 1}{3} = \frac{7}{3}$

To change the improper fraction to the mixed number, divide the numerator by the denominator.

The partial quotient is the whole number, the remainder is the numerator, and the denominator remains the same:

$\frac{7}{4} = 1\frac{3}{4}$ $\frac{24}{6} = 4$

Main Property of Fraction

If the numerator and denominator of a fraction multiplied or divided by the same natural number, to the new fraction will equal the first fraction:

$\frac{1}{2} = \frac{2}{4}$ $\frac{2}{4} = \frac{1}{2}$

Two equal fractions can have different notations of the same number:

$\frac{a}{b} = \frac{a \cdot n}{b \cdot n}$ $\frac{1}{2} = \frac{1 \cdot 2}{2 \cdot 2} = \frac{1}{4}$

Grade 7 Chapter 4 Lesson 2

Find the missing numerator:

$\frac{4}{5}=\frac{?}{25}$ $\quad 25 \div 5 = 5 \quad 4 \times 5 = 20 \quad$ $\frac{4}{5}=\frac{4\times5}{5\times5}=\frac{20}{25}$

To reduce the fraction means to divide the numerator and denominator of the fraction by the same natural number:

$\frac{a\div n}{b\div n}=\frac{a}{b}$ $\qquad$ $\frac{2}{4}=\frac{2\div2}{4\div2}=\frac{1}{2}$

If the GCF of the numerator and denominator equal one, the fraction cannot reduced.

If the GCF (5, 6) = 1, fraction $\frac{5}{6}$ cannot be reduced.

Greatest Common Factor (GCF)

The greatest number by which a fraction can reduced is the greatest common factor of the numerator and denominator:

GC F (12, 18) = 6

$\frac{12}{18}=\frac{12\div6}{18\div6}=\frac{2}{3}$ or $\frac{12}{18}=\frac{12\div2}{18\div2}=\frac{6}{9}=\frac{6\div3}{9\div3}=\frac{2}{3}$

To easily reduce a fraction, divide by small numbers 2, 3, 4, 5, …

$\frac{60}{75}=\frac{60\div3}{75\div3}=\frac{20}{25}=\frac{20\div5}{25\div5}=\frac{4}{5}$

$\frac{360}{600}=\frac{360\div10}{600\div10}=\frac{36}{60}=\frac{36\div2}{60\div2}=\frac{18}{30}=\frac{18\div2}{30\div2}=\frac{9}{15}=\frac{9\div3}{15\div3}=\frac{3}{5}$

Reduce a fraction by using prime factorization:

$\frac{360}{600}=\frac{2\times2\times2\times3\times3\times5}{2\times2\times2\times3\times5\times5}=\frac{3}{5}$

To work with fractions let us refresh our memory of GCF and LCM from Table 4

Grade 7 Chapter 4 Lesson 2

Table 4

Greatest Common Factor (in short GCF)

The Greatest Common Factor is the greatest number by which all given numbers can be divided.

1) Find the GCF of numbers 12 and 18 (÷ 1, 2, 3, 4, ...)

Divisors of number 12 are 1, 2, 3, 4, 6, 12

Divisors of number 18 are 1, 2, 3, 6, 9, 18

Common divisors are 1, 2, 3, 6 Greatest common divisor is 6

GCD (12, 18) = 6 $12 \div 6 = 2$ $18 \div 6 = 3$

2)If greatest of the numbers can divided evenly by the smallest number, then the greatest common factor of numbers is equal to the smallest of given numbers: Find the GCF of numbers 12 and 4

Factors of 12 are 1, 2, 3, 4, 6, 12 Factors of 4 are 1, 2, 4

Common factors are 1, 2, 4 Greatest common factor is 4

GCF (12, 4) = 4 $12 \div 4 = 3$

3) The greatest common factor of prime numbers equal one: GCF (5,7) = 1

Least Common Multiples (in short LCM)

The least number is the smallest number that can be evenly divided by each of the given numbers

1) Find the LCM of numbers 12 and 18 (×1, 2, 3, 4, ...)

Multiples of 12 are 12, 24, 36, 48, 60, 72...

Multiples of 18 are 18, 36, 54, 72...

Common multiples are 36, 72... Least common multiples is 36

LCM (12, 18) = 36 $36 \div 12 = 3$ $36 \div 18 = 2$

2)If the greatest of given numbers evenly divided by the smallest number, the LCM of the numbers is equal to the greatest of the given numbers: Find the LCM of numbers 4, 12

Multiples of 4 are 4, 8, 12, 16, 20, 24... Multiples of 12 are 12, 24...

Common multiples are 12, 24... LCM (4, 12) = 12

3)The least common multiple of prime numbers is equal to the product of the numbers. Find the LCM of numbers 2 and 5

Multiples of 2 are 2, 4, 6, 8, 10... Multiples of 5 are 5, 10...

Common multiples are 10, 20 ... LCM (2, 5) = $2 \times 5 = 10$

Grade 7 Chapter 4 Lesson 3
Comparison of Fractions and Mixed numbers

To compare two fractions with the same denominators, compare the numerators.

The greater fraction has the greater numerator.
The smaller fraction has the smaller numerator:

$\frac{4}{5} > \frac{2}{5}$ $4 > 2$ $\frac{1}{7} < \frac{6}{7}$ $1 < 6$

To compare two fractions with the same numerators, compare the denominators.

The greater fraction has the smaller denominator.
The smaller fraction has the greater denominator:

$\frac{1}{2} > \frac{1}{3}$ $2 < 3$ $\frac{1}{4} < \frac{1}{3}$ $4 > 3$

Find the Least Common Denominator (LCD) of fractions $\frac{1}{4}$ and $\frac{3}{8}$

If the greatest denominator can be divided by the smallest denominator, the greatest denominator is LCD $8 \div 4 = 2$
LCD (4, 8) = 8

The number needed to multiply the numerator and denominator of a fraction to get a new fraction called the factor.

To find the factor of each fraction, divide the least common denominator (LCD) by each denominator: $8 \div 4 = 2$ $8 \div 8 = 1$

Multiply the numerator and denominator of each fraction by the factor to get fractions with the same denominators:

$\frac{1}{4} = \frac{1 \cdot 2}{4 \cdot 2} = \frac{2}{8}$ $\frac{3}{8} = \frac{3 \times 1}{8 \times 1} = \frac{3}{8}$ $\frac{2}{8}$ and $\frac{3}{8}$

Grade 7 Chapter 4 Lesson 3
Compare Fractions and Mixed Numbers with Different Denominators

To compare two fractions with different denominators, find the least common denominator and compare numerators: $1\frac{5}{6}$ **and** $1\frac{1}{8}$

To find least common denominators of fractions (LCD), find the least common multiple of each denominator: LCM (6, 8).

If the greatest denominator can divide by the smallest denominator, the greatest denominator is LCD

If the greater denominator cannot be divided by the smaller denominator, multiply the greater denominator by 2, 3, 4, … until the new denominator can divided evenly, (remainder zero) by the smaller denominator.

Example: Compare Mixed Numbers: $1\frac{5}{6}$ **and** $1\frac{1}{8}$

Find the LCM (6, 8)   **8 cannot divided evenly by 6**

$8\times2 = 16$ **16 cannot be divided evenly by 6**

$8\times3=24$ **24 can be divided evenly by 8 and by 6**   **LCD (6, 8) = 24**

$24 \div 6 = 4$ $24 \div 8 = 3$ $\frac{5}{6} = \frac{5\cdot4}{6\cdot4} = \frac{20}{24}$ $\frac{1}{8} = \frac{1\cdot3}{8\cdot3} = \frac{3}{24}$

$\frac{20}{24} > \frac{3}{24}$ $\frac{5}{6} > \frac{1}{8}$ $1\frac{5}{6} > 1\frac{1}{8}$

Second way $\frac{5}{6} \; ? \; \frac{1}{8}$

Find cross product 5×8 and 6×1 40 and 6

But $40 > 6$ $\frac{5}{6} > \frac{1}{8}$

Grade 7 Chapter 4 Lesson 3
Review 1 A

Represent answers like fractions

1) $5 \div 6$ 2) $7 \div 8$ 3) $9 \div 9$ 4) $9 \div 8$

Compare numbers (put <, >, =)

5) $\frac{5}{6} ? 1$ 6) $\frac{7}{8} ? 1$ 7) $\frac{9}{9} ? 1$ 8) $\frac{9}{8} ? 1$

Add numbers

9) $2 + \frac{3}{4}$ 10) $\frac{3}{4} + 3$ 11) $4 + \frac{4}{5}$ 12) $\frac{6}{7} + 5$

Change mixed number in the improper fraction

13) $2\frac{3}{4}$ 14) $4\frac{3}{5}$ 15) $5\frac{5}{6}$ 16) $6\frac{1}{8}$

Change improper fraction in the mixed number

17) $\frac{13}{4}$ 18) $\frac{22}{5}$ 19) $\frac{35}{6}$ 20) $\frac{53}{8}$

Find the missing numerator

21) $\frac{5}{6} = \frac{?}{12}$ 22) $\frac{6}{7} = \frac{?}{21}$ 23) $\frac{7}{8} = \frac{?}{32}$ 24) $\frac{8}{9} = \frac{?}{45}$

Reduce fraction

25) $\frac{4}{10}$ 26) $\frac{9}{12}$ 27) $\frac{9}{15}$ 28) $\frac{16}{20}$

29) $\frac{18}{24}$ 30) $\frac{20}{30}$ 31) $\frac{25}{30}$ 32) $\frac{24}{36}$

33) $\frac{16}{40}$ 34) $\frac{20}{40}$ 35) $\frac{25}{45}$ 36) $\frac{15}{50}$

37) $\frac{45}{60}$ 38) $\frac{48}{80}$ 39) $\frac{72}{90}$ 40) $\frac{75}{100}$

Grade 7 Chapter 4 Lesson 3
Review 1 B
Put correct symbols <, >, = instead question mark

1) $\frac{1}{3}$? $\frac{2}{6}$	2) $\frac{3}{6}$? $\frac{1}{2}$	3) $\frac{2}{3}$? $\frac{4}{6}$	4) $\frac{6}{8}$? $\frac{3}{4}$
5) $\frac{1}{3}$? $\frac{2}{3}$	6) $\frac{2}{4}$? $\frac{1}{4}$	7) $\frac{2}{7}$? $\frac{4}{7}$	8) $\frac{5}{8}$? $\frac{3}{8}$
9) $\frac{1}{4}$? $\frac{1}{3}$	10) $\frac{2}{3}$? $\frac{2}{5}$	11) $\frac{3}{7}$? $\frac{3}{4}$	12) $\frac{5}{6}$? $\frac{5}{8}$
13) $\frac{1}{2}$? $\frac{3}{4}$	14) $\frac{4}{6}$? $\frac{1}{2}$	15) $\frac{1}{3}$? $\frac{4}{9}$	16) $\frac{3}{8}$? $\frac{1}{4}$
17) $\frac{2}{3}$? $\frac{5}{6}$	18) $\frac{7}{8}$? $\frac{3}{4}$	19) $\frac{4}{5}$? $\frac{9}{10}$	20) $\frac{11}{12}$? $\frac{5}{6}$
21) $\frac{2}{5}$? $\frac{1}{3}$	22) $\frac{1}{7}$? $\frac{2}{3}$	23) $\frac{3}{4}$? $\frac{1}{3}$	24) $\frac{1}{6}$? $\frac{2}{7}$
25) $\frac{3}{4}$? $\frac{5}{6}$	26) $\frac{5}{6}$? $\frac{3}{8}$	27) $\frac{5}{6}$? $\frac{8}{9}$	28) $\frac{7}{8}$? $\frac{5}{12}$
29) $\frac{2}{3}$? $\frac{4}{6}$	30) $\frac{5}{6}$? $\frac{1}{2}$	31) $\frac{2}{3}$? $\frac{5}{6}$	32) $\frac{3}{4}$? $\frac{6}{8}$
33) $\frac{3}{4}$? $\frac{4}{8}$	34) $\frac{4}{5}$? $\frac{4}{10}$	35) $\frac{1}{4}$? $\frac{2}{8}$	36) $\frac{1}{3}$? $\frac{3}{9}$
37) $1\frac{1}{6}$? $1\frac{1}{3}$	38) $2\frac{1}{3}$? $2\frac{1}{6}$	39) $1\frac{2}{4}$? $1\frac{1}{2}$	40) $2\frac{3}{4}$? $2\frac{1}{2}$

Grade 7 Chapter 4 Lesson 4

Addition and Subtraction of Fractions and Mixed Numbers

Addition of Fractions and Mixed numbers

The Same Denominators

To add fractions with the same denominators, add their numerators; the denominator remains the same

$$\frac{a}{c}+\frac{b}{c}=\frac{a+b}{c}$$

$$1=\frac{a}{a}=\frac{2}{2}=\frac{3}{3}=\frac{4}{4}=\frac{5}{5}$$

a) $\frac{4}{8}+\frac{3}{8}+\frac{1}{8}=\frac{4+3+1}{8}=\frac{8}{8}=1$

b) $\frac{1}{20}+\frac{3}{20}+\frac{7}{20}+\frac{9}{20}=\frac{1+3+7+9}{20}=\frac{20}{20}=1$

Different denominators

To add fractions with different denominators, find the least common denominator, multiply the fractions by a factor and add the numerators; the denominator remains the same.

LCD (6, 12) = 12 **LCD (5, 8, 20) = 40**

c) $\frac{1}{6}+\frac{7}{12}=\frac{1\cdot 2}{6\cdot 2}+\frac{7}{12}=\frac{2}{12}+\frac{7}{12}=\frac{9}{12}=\frac{9\div 3}{12\div 3}=\frac{3}{4}$

d) $\frac{2}{5}+\frac{3}{8}+\frac{7}{20}=\frac{2\cdot 8}{5\cdot 8}+\frac{3\cdot 5}{8\cdot 5}+\frac{7\cdot 2}{20\cdot 2}=\frac{16}{40}+\frac{15}{40}+\frac{14}{40}=\frac{45}{40}=\frac{45\div 5}{40\div 5}=\frac{9}{8}=1\frac{1}{8}$

Mixed numbers

To add mixed numbers with different denominators, find the least common denominator, add whole numbers and add numerators, the denominator remains the same.

In addition, mixed numbers can change to improper fractions and then fractions can add.

Remember: $1=\frac{a}{a}=\frac{2}{2}=\frac{3}{3}=\frac{4}{4}=\frac{5}{5}$ $1=\frac{10}{10}$ $\frac{14}{12}=1\frac{2}{12}$

Grade 7 Chapter 4 Lesson 4
The Same Denominators

a) $1\frac{1}{8}+2\frac{3}{8}=3\frac{4}{8}=3\frac{1}{2}$

b) $1\frac{1}{9}+2\frac{2}{9}+4\frac{3}{9}=7\frac{6}{9}=7\frac{2}{3}$

c) $2\frac{3}{10}+1\frac{7}{10}=3\frac{10}{10}=3+\frac{10}{10}=3+1=4$

d) $1\frac{5}{12}+2\frac{9}{12}=3\frac{14}{12}=3+\frac{14}{12}=3+1\frac{2}{12}=4\frac{2}{12}=4\frac{1}{6}$

Different Denominators

e) $1\frac{1}{3}+2\frac{3}{9}=1\frac{1}{3}+2\frac{3\div 3}{9\div 3}=1\frac{1}{3}+2\frac{1}{3}=3\frac{2}{3}$

f) $1\frac{2}{6}+2\frac{1}{9}=1\frac{2\div 2}{6\div 2}+2\frac{1}{9}=1\frac{1}{3}+2\frac{1}{9}=1\frac{1\times 3}{3\times 3}+2\frac{1}{9}=1\frac{3}{9}+2\frac{1}{9}=3\frac{4}{9}$

g) $1\frac{1}{2}+1\frac{1}{3}+2\frac{1}{6}=1\frac{1\cdot 3}{2\cdot 3}+1\frac{1\cdot 2}{3\cdot 2}+2\frac{1}{6}=1\frac{3}{6}+1\frac{2}{6}+2\frac{1}{6}=4\frac{6}{6}=5$

LCD (2, 3, 6) = 6

Change mixed numbers to improper fractions:

h) $1\frac{1}{2}+1\frac{1}{3}+2\frac{1}{6}=\frac{3}{2}+\frac{4}{3}+\frac{13}{6}=\frac{3\cdot 3}{2\cdot 3}+\frac{4\cdot 2}{3\cdot 2}+\frac{13}{6}=\frac{9}{6}+\frac{8}{6}+\frac{13}{6}=\frac{30}{6}=5$

Grade 7 Chapter 4 Lesson 4
Subtraction Fractions and Mixed Numbers

To subtract fractions with the same denominator, subtract the numerators; the denominator remains the same.
Always first, try to reduce fractions:

$$\frac{a}{c} - \frac{b}{c} = \frac{a-b}{c}$$

a) $\frac{5}{8} - \frac{3}{8} = \frac{5-3}{8} = \frac{2}{8} = \frac{1}{4}$

To subtract fractions with different denominators, expand fractions to the least common denominator and subtract numerators; the denominator remains the same.

LCD (10, 4) = 20 **LCD (12, 18) = 36**

b) $\frac{6}{20} - \frac{2}{8} = \frac{3}{10} - \frac{1}{4} = \frac{6}{20} - \frac{5}{20} = \frac{1}{20}$

c) $\frac{11}{12} - \frac{3}{18} = \frac{11}{12} - \frac{1}{6} = \frac{11-2}{12} = \frac{9}{12} = \frac{3}{4}$

To subtract mixed numbers, expand fractions to the least common denominator, subtract whole numbers and subtract numerators; the denominator remains the same.

Also, mixed numbers can be changed to improper fractions and then subtract fractions.

Remember: $4 = 3\frac{8}{8}$

$8 = 7 + 1 = 7 + \frac{6}{6} = 7\frac{6}{6}$

d) $6\frac{7}{8} - 5\frac{3}{8} = 1\frac{4}{8} = 1\frac{1}{2}$

e) $1 - \frac{2}{3} = \frac{3}{3} - \frac{2}{3} = \frac{1}{3}$

f) $8 - 3\frac{5}{6} = 7\frac{6}{6} - 3\frac{5}{6} = 4\frac{1}{6}$

g) $3\frac{1}{2} - 2 = 1\frac{1}{2}$

h) $4 - \frac{5}{8} = 3\frac{8}{8} - \frac{5}{8} = 3\frac{3}{8}$

i) $3\frac{3}{4} - 2\frac{1}{2} = \frac{15}{4} - \frac{5}{2} = \frac{15}{4} - \frac{10}{4} = \frac{5}{4} = 1\frac{1}{4}$

Grade 7 Chapter 4 Lesson 4
Special Case of Subtraction

If the numerator of the minuend is less than the numerator of the subtrahend, the whole numbers of the minuend need to decrease by 1.

To get the new numerator of the minuend, add the old numerator to the denominator.

Subtract the whole numbers and numerators; the denominator remains the same.

a) $6\frac{2}{7} - 2\frac{5}{7} = 5\frac{9}{7} - 2\frac{5}{7} = 3\frac{4}{7}$

$(6\frac{2}{7} = 5+1+\frac{2}{7} = 5+1\frac{2}{7} = 5+\frac{9}{7} = 5\frac{9}{7})$

b) $5\frac{1}{3} - 2\frac{5}{6} = 5\frac{2}{6} - 2\frac{5}{6} = 4\frac{8}{6} - 2\frac{5}{6} = 2\frac{3}{6} = 2\frac{1}{2}$

$(5\frac{2}{6} = 4\frac{8}{6}$ $\quad 5 - 1 = 4 \quad$ $2 + 6 = 8)$

c) $6\frac{1}{4} - 2\frac{2}{3} = 6\frac{3}{12} - 2\frac{8}{12} = 5\frac{15}{12} - 2\frac{8}{12} = 3\frac{7}{12}$

$\left(6\frac{3}{12} = 5+1+\frac{3}{12} = 5+\frac{12}{12}+\frac{3}{12} = 5\frac{15}{12}\right)$

d) $6\frac{1}{4} - 2\frac{2}{3} = \frac{25}{4} - \frac{8}{3} = \frac{75}{12} - \frac{32}{12} = \frac{43}{12} = 3\frac{7}{12}$

Grade 7 Chapter 4 Lesson 4
Review 1 A
Add the fractions and mixed numbers

1) $\frac{1}{2}+\frac{1}{2}$	2) $\frac{1}{3}+\frac{2}{3}$	3) $\frac{1}{4}+\frac{1}{4}$	4) $\frac{1}{4}+\frac{3}{4}$
5) $\frac{1}{5}+\frac{4}{5}$	6) $\frac{1}{6}+\frac{2}{6}$	7) $\frac{1}{6}+\frac{3}{6}$	8) $\frac{1}{6}+\frac{5}{6}$
9) $\frac{1}{8}+\frac{3}{8}$	10) $\frac{1}{8}+\frac{5}{8}$	11) $\frac{1}{3}+\frac{3}{9}$	12) $\frac{1}{3}+\frac{1}{6}$
13) $\frac{1}{4}+\frac{2}{8}$	14) $\frac{2}{3}+\frac{4}{6}$	15) $\frac{1}{2}+\frac{3}{4}$	16) $\frac{3}{4}+\frac{4}{8}$
17) $\frac{4}{5}+\frac{4}{10}$	18) $\frac{5}{6}+\frac{1}{2}$	19) $\frac{2}{3}+\frac{5}{6}$	20) $\frac{3}{4}+\frac{6}{8}$
21) $3\frac{2}{4}+1\frac{1}{2}$	22) $2\frac{3}{4}+1\frac{1}{2}$	23) $1\frac{1}{6}+2\frac{1}{3}$	24) $2\frac{1}{6}+1\frac{1}{3}$
25) $1\frac{3}{4}+3\frac{1}{2}$	26) $1\frac{2}{3}+2\frac{1}{2}$	27) $5\frac{2}{3}+1\frac{6}{9}$	28) $1\frac{4}{8}+5\frac{3}{4}$
29) $4\frac{2}{16}+2\frac{5}{8}$	30) $3\frac{4}{12}+2\frac{5}{6}$	31) $2\frac{1}{14}+3\frac{3}{7}$	32) $2\frac{7}{18}+4\frac{1}{9}$
33) $1\frac{1}{2}+2\frac{3}{9}$	34) $1\frac{3}{6}+2\frac{1}{8}$	35) $2\frac{2}{4}+1\frac{1}{3}$	36) $4\frac{3}{12}+2\frac{3}{8}$
37) $1\frac{2}{4}+3\frac{2}{5}$	38) $3\frac{1}{6}+1\frac{3}{9}$	39) $1\frac{3}{6}+4\frac{3}{4}$	40) $4\frac{2}{4}+1\frac{5}{6}$

Grade 7 Chapter 4 Lesson 4
Review 1 B
Subtract the fractions and the mixed numbers

1) $\frac{2}{3} - \frac{1}{3}$ 2) $\frac{3}{4} - \frac{1}{4}$ 3) $\frac{5}{7} - \frac{3}{7}$ 4) $\frac{7}{8} - \frac{3}{8}$

5) $1\frac{1}{6} - \frac{1}{6}$ 6) $1 - \frac{3}{4}$ 7) $1 - \frac{2}{3}$ 8) $1\frac{1}{6} - \frac{5}{6}$

9) $1\frac{1}{6} - \frac{3}{6}$ 10) $1\frac{7}{8} - \frac{5}{8}$ 11) $1\frac{1}{3} - \frac{2}{3}$ 12) $1\frac{1}{2} - \frac{3}{4}$

13) $1\frac{1}{4} - \frac{1}{2}$ 14) $1\frac{1}{6} - \frac{1}{2}$ 15) $1\frac{7}{9} - \frac{1}{3}$ 16) $1\frac{5}{8} - \frac{1}{4}$

17) $1\frac{1}{2} - \frac{5}{6}$ 18) $1\frac{5}{8} - \frac{3}{4}$ 19) $1\frac{7}{10} - \frac{4}{5}$ 20) $1\frac{3}{4} - \frac{11}{12}$

21) $3\frac{3}{9} - 2\frac{1}{3}$ 22) $6\frac{3}{6} - 5\frac{3}{4}$ 23) $2\frac{1}{3} - 1\frac{1}{6}$ 24) $5\frac{4}{8} - 4\frac{1}{6}$

25) $2\frac{6}{12} - 1\frac{4}{8}$ 26) $3\frac{6}{9} - 2\frac{1}{6}$ 27) $4\frac{3}{6} - 3\frac{7}{14}$ 28) $2\frac{5}{6} - 1\frac{2}{8}$

29) $4\frac{6}{16} - 3\frac{2}{8}$ 30) $6\frac{6}{9} - 5\frac{9}{18}$ 31) $4\frac{1}{4} - 3\frac{3}{4}$ 32) $2\frac{4}{8} - 1\frac{4}{12}$

33) $6\frac{5}{6} - 5\frac{3}{4}$ 34) $5\frac{1}{4} - 4\frac{3}{6}$ 35) $5\frac{6}{8} - 4\frac{4}{16}$ 36) $1\frac{4}{8} - 1\frac{5}{12}$

37) $3\frac{3}{4} - 2\frac{3}{6}$ 38) $2\frac{3}{6} - 2\frac{1}{9}$ 39) $3\frac{5}{6} - 2\frac{7}{18}$ 40) $2\frac{7}{8} - 1\frac{7}{24}$

Grade 7 Chapter 4 Lesson 5

Multiplication and Powers Fractions and Mixed Numbers

Multiplication of the Fractions

To multiply the fractions, product of numerators put in new numerator and product of denominators put in new denominator.

Always first try to reduce fractions $\frac{a}{b}\times\frac{c}{d}=\frac{a\times c}{b\times d}=\frac{ac}{bd}$

a) $\frac{5}{6}\times\frac{3}{10}=\frac{5\times3}{6\times10}=\frac{1\times1}{2\times2}=\frac{1}{4}$ b) $\frac{3}{5}\times\frac{10}{11}\times\frac{22}{30}=\frac{3\times10\times22}{5\times11\times30}=\frac{1\times2\times2}{1\times1\times10}=\frac{2}{5}$

Multiplication of the Fraction and Natural Number

To multiply the fractions by natural number, multiply numerator by natural number, denominator remains the same. $a=\frac{a}{1}$ $8=\frac{8}{1}$

$\frac{a}{b}\cdot c=\frac{a}{b}\cdot\frac{c}{1}=\frac{a\cdot c}{b}$ $a\cdot\frac{b}{c}=\frac{a}{1}\cdot\frac{b}{c}=\frac{ab}{c}$

c) $\frac{3}{4}\times8=\frac{3}{4}\times\frac{8}{1}=\frac{3}{1}\times\frac{2}{1}=\frac{6}{1}=6$ d) $8\times\frac{5}{2}=8\div2\times5=4\times5=20$

Multiplication of Mixed Numbers

To multiply the mixed numbers, change the mixed numbers to improper fractions and multiply fractions.

e) $3\frac{1}{3}\cdot2\frac{2}{5}=\frac{10}{3}\cdot\frac{12}{5}=\frac{2\cdot4}{1\cdot1}=8$ f) $\frac{5}{8}\cdot4\frac{4}{5}\cdot\frac{1}{6}=\frac{5}{8}\cdot\frac{24}{5}\cdot\frac{1}{6}=\frac{3}{6}=\frac{1}{2}$

Power of Mixed Number

To find the power of a mixed numbers, change the mixed number to an improper fraction, then find the power of the numerator and denominator:

$\left(\frac{a}{b}\right)^n=\frac{a^n}{b^n}$ $\left(\frac{1}{2}\right)^2=\frac{1^2}{2^2}=\frac{1}{4}$

g) $\left(\frac{3}{4}\right)^3=\frac{3^3}{4^3}=\frac{27}{64}$ h) $\left(1\frac{1}{2}\right)^2=\left(\frac{3}{2}\right)^2=\frac{3^2}{2^2}=\frac{9}{4}=2\frac{1}{4}$

Grade 7 Chapter 4 Lesson 5
Division of Fractions and Mixed numbers
Reciprocals Fractions

The reciprocal of a fraction is the fraction with the numerator and denominator interchanged.
Product of two reciprocals fractions is equal one.

$a\cdot\frac{1}{a}=1$ $\frac{a}{b}\cdot\frac{b}{a}=1$ $a)\,5\cdot\frac{1}{5}=\frac{5}{1}\cdot\frac{1}{5}=1$

$b)\,\frac{1}{3}\cdot 3=\frac{1}{3}\cdot\frac{3}{1}=1$ $c)\,\frac{2}{3}\cdot\frac{3}{2}=\frac{2\cdot 3}{3\cdot 2}=1$ $d)\,1\frac{3}{4}\cdot\frac{4}{7}=\frac{7}{4}\cdot\frac{4}{7}=1$

Division of Fractions

To divide fractions, dividend multiplies by number, reciprocal of divisor. In multiplication, always first try to reduce the fraction

$\frac{a}{b}\div\frac{c}{d}=\frac{a}{b}\cdot\frac{d}{c}=\frac{ad}{dc}$ $e)\,\frac{3}{4}\div\frac{3}{8}=\frac{3}{4}\cdot\frac{8}{3}=\frac{1\times 2}{1\times 1}=\frac{2}{1}=2$

$\frac{a}{b}\div c=\frac{a}{b}\cdot\frac{1}{c}=\frac{a}{bc}$ $f)\,\frac{3}{4}\div 9=\frac{3}{4}\div\frac{9}{1}=\frac{3}{4}\cdot\frac{1}{9}=\frac{1}{12}$

$a\div\frac{b}{c}=\frac{a}{1}\cdot\frac{c}{b}=\frac{ac}{b}$ $g)\,9\div\frac{3}{4}=9\times\frac{4}{3}=9\div 3\times 4=3\times 4=12$

Division of Mixed Numbers

To divided mixed numbers, change mixed numbers to improper fractions and divide the fractions

$h)\,1\frac{1}{6}\div 4\frac{2}{3}=\frac{7}{6}\div\frac{14}{3}=\frac{7}{6}\times\frac{3}{14}=\frac{1}{2}\times\frac{1}{2}=\frac{1}{4}$

$i)\,3\frac{1}{5}\div 2\frac{2}{5}=\frac{16}{5}\div\frac{12}{5}=\frac{16}{5}\times\frac{5}{12}=\frac{4}{1}\times\frac{1}{3}=\frac{4}{3}=1\frac{1}{3}$

$j)\,8\div 4\frac{4}{5}=\frac{8}{1}\div\frac{24}{5}=\frac{8}{1}\times\frac{5}{24}=\frac{1}{1}\times\frac{5}{3}=\frac{5}{3}=1\frac{2}{3}$

$l)\,4\frac{1}{3}\div 13=\frac{13}{3}\div\frac{13}{1}=\frac{13}{3}\times\frac{1}{13}=\frac{1}{3}$

Grade 7 Chapter 4 Lesson 5
Review 1 A
Multiply the fractions and the mixed numbers

1) $\frac{2}{3}\times\frac{3}{4}$ 2) $\frac{3}{4}\times\frac{4}{5}$ 3) $\frac{4}{5}\times\frac{5}{6}$ 4) $\frac{3}{4}\times\frac{6}{9}$

5) $\frac{2}{3}\times\frac{3}{5}$ 6) $\frac{3}{5}\times\frac{5}{9}$ 7) $\frac{4}{5}\times\frac{15}{16}$ 8) $\frac{5}{6}\times\frac{18}{25}$

9) $\frac{2}{3}\times 6$ 10) $\frac{3}{4}\times 12$ 11) $15\times\frac{4}{5}$ 12) $18\times\frac{5}{6}$

13) $3\times\frac{1}{6}$ 14) $4\times\frac{5}{12}$ 15) $\frac{7}{15}\times 5$ 16) $\frac{9}{18}\times 6$

17) $2\frac{1}{2}\times 3\frac{1}{3}$ 18) $4\frac{1}{2}\times 2\frac{2}{3}$ 19) $2\frac{1}{3}\times 1\frac{2}{7}$ 20) $1\frac{3}{5}\times 2\frac{2}{4}$

Find exponent of number

21) $\left(\frac{1}{2}\right)^2$ 22) $\left(\frac{1}{3}\right)^2$ 23) $\left(\frac{1}{4}\right)^2$ 24) $\left(\frac{1}{5}\right)^2$

25) $\left(\frac{2}{3}\right)^2$ 26) $\left(\frac{2}{5}\right)^2$ 27) $\left(\frac{4}{5}\right)^2$ 28) $\left(\frac{5}{6}\right)^2$

29) $\left(\frac{2}{4}\right)^2$ 30) $\left(\frac{4}{6}\right)^2$ 31) $\left(\frac{1}{2}\right)^3$ 32) $\left(\frac{2}{3}\right)^3$

33) $\left(1\frac{1}{2}\right)^2$ 34) $\left(1\frac{1}{3}\right)^2$ 35) $\left(1\frac{2}{3}\right)^2$ 36) $\left(1\frac{1}{4}\right)^2$

37) $\left(1\frac{2}{4}\right)^2$ 38) $\left(1\frac{3}{9}\right)^2$ 39) $\left(1\frac{4}{6}\right)^2$ 40) $\left(1\frac{2}{8}\right)^2$

Grade 7 Chapter 4 Lesson 5
Review 1 B
Divide the fractions and the mixed numbers

1) $\frac{3}{2} \div \frac{1}{6}$	2) $\frac{2}{3} \div \frac{1}{6}$	3) $\frac{3}{4} \div \frac{1}{8}$	4) $\frac{7}{15} \div \frac{1}{5}$
5) $\frac{5}{18} \div \frac{1}{6}$	6) $\frac{2}{3} \div \frac{1}{3}$	7) $\frac{1}{2} \div \frac{1}{4}$	8) $4 \div \frac{1}{2}$
9) $\frac{1}{2} \div \frac{3}{4}$	10) $\frac{1}{3} \div \frac{4}{9}$	11) $\frac{2}{3} \div \frac{5}{9}$	12) $\frac{3}{4} \div \frac{9}{10}$
13) $\frac{5}{6} \div 5$	14) $\frac{6}{7} \div 6$	15) $\frac{7}{8} \div 7$	16) $\frac{8}{9} \div 8$
17) $9 \div \frac{9}{10}$	18) $5 \div \frac{5}{6}$	19) $10 \div \frac{5}{6}$	20) $6 \div \frac{3}{4}$
21) $\frac{1}{3} \div 1\frac{1}{3}$	22) $\frac{1}{4} \div 1\frac{1}{4}$	23) $\frac{1}{5} \div 1\frac{1}{5}$	24) $\frac{3}{5} \div 1\frac{4}{5}$
25) $1\frac{1}{5} \div \frac{4}{5}$	26) $1\frac{1}{6} \div \frac{5}{6}$	27) $1\frac{1}{4} \div 5$	28) $1\frac{1}{6} \div 7$
29) $2\frac{1}{2} \div 1\frac{1}{4}$	30) $3\frac{3}{4} \div 1\frac{1}{4}$	31) $4\frac{1}{5} \div 1\frac{2}{5}$	32) $2\frac{2}{3} \div 1\frac{1}{3}$
33) $1\frac{1}{2} \div 1\frac{1}{2}$	34) $2\frac{2}{3} \div 2\frac{2}{3}$	35) $3\frac{1}{4} \div 3\frac{1}{4}$	36) $2\frac{1}{5} \div 2\frac{1}{5}$
37) $2\frac{1}{2} \div 1\frac{1}{2}$	38) $3\frac{2}{3} \div 1\frac{1}{3}$	39) $3\frac{3}{4} \div 2\frac{2}{8}$	40) $4\frac{4}{5} \div 3\frac{2}{10}$

Grade 7 Chapter 4 Lesson 6

Table 5 Fractions

Properties of Fraction

$$\frac{a}{b}=\frac{a\cdot n}{b\cdot n}$$

$$\frac{a\div n}{b\div n}=\frac{a}{b}$$

$$\frac{2}{3}=\frac{2\times 6}{3\times 6}=\frac{12}{18}$$

$$\frac{60}{75}=\frac{60\div 3}{75\div 3}=\frac{20}{25}=\frac{20\div 5}{25\div 5}=\frac{4}{5}$$

Comparison of Fractions

$$\frac{6}{11}>\frac{7}{22}$$

$$\frac{6}{11}=\frac{6\cdot 2}{11\cdot 2}=\frac{12}{22}$$

$$\frac{12}{22}>\frac{7}{22}$$

Addition of Fractions

$$\frac{a}{c}+\frac{b}{c}=\frac{a+b}{c}$$

$$\frac{2}{5}+\frac{3}{8}+\frac{7}{20}=\frac{2\cdot 8}{5\cdot 8}+\frac{3\cdot 5}{8\cdot 5}+\frac{7\cdot 2}{20\cdot 2}=\frac{16}{40}+\frac{15}{40}+\frac{14}{40}=\frac{45}{40}=1\frac{1}{8}$$

Subtraction of Fractions

$$\frac{a}{c}-\frac{b}{c}=\frac{a-b}{c}$$

$$\frac{3}{10}-\frac{1}{4}=\frac{6}{20}-\frac{5}{20}=\frac{1}{20}$$

$$\frac{1}{12}-\frac{1}{18}=\frac{3}{36}-\frac{2}{36}=\frac{3-2}{36}=\frac{1}{36}$$

Multiplication of Fractions

$$\frac{a}{b}\times\frac{c}{d}=\frac{a\times c}{b\times d}=\frac{ac}{bd}$$

$$\frac{4}{6}\times\frac{9}{8}=\frac{4\times 9}{6\times 8}=\frac{1\times 3}{2\times 2}=\frac{3}{4}$$

$$\frac{a}{b}\cdot c=\frac{a}{b}\cdot\frac{c}{1}=\frac{a\cdot c}{b}$$

$$\frac{3}{8}\times 4=\frac{3}{8}\times\frac{4}{1}=\frac{3\times 1}{2\times 1}=\frac{3}{2}=1\frac{1}{2}$$

$$a\cdot\frac{b}{c}=\frac{a}{1}\cdot\frac{b}{c}=\frac{ab}{c}$$

$$16\times\frac{5}{8}=\frac{16}{1}\times\frac{5}{8}=\frac{2\times 5}{1}=10$$

Division of Fractions

$$\frac{a}{b}\div\frac{c}{d}=\frac{a}{b}\cdot\frac{d}{c}=\frac{ad}{dc}$$

$$\frac{5}{6}\div\frac{10}{3}=\frac{5}{6}\times\frac{3}{10}=\frac{1\times 1}{2\times 2}=\frac{1}{4}$$

$$\frac{a}{b}\div c=\frac{a}{b}\cdot\frac{1}{c}=\frac{a}{bc}$$

$$\frac{3}{4}\div 6=\frac{3}{4}\div\frac{6}{1}=\frac{3}{4}\times\frac{1}{6}=\frac{1}{8}$$

$$a\div\frac{b}{c}=\frac{a}{1}\cdot\frac{c}{b}=\frac{ac}{b}$$

$$6\div\frac{3}{4}=\frac{6}{1}\times\frac{4}{3}=\frac{8}{1}=8$$

Power of Fractions

$$\left(\frac{a}{b}\right)^n=\frac{a^n}{b^n}$$

$$\left(\frac{1}{3}\right)^2=\frac{1^2}{3^2}=\frac{1}{9}$$

Grade 7 Chapter 4 Lesson 6

Table 6 Mixed Numbers

Mixed Number $2+\frac{1}{2}=2\frac{1}{2}$ $\quad \frac{1}{4}+1=1\frac{1}{4}$ $\quad 1\frac{3}{4}=\frac{4\cdot 1+3}{4}=\frac{7}{4}$

Comparison of Mixed Numbers $1\frac{1}{3}<1\frac{5}{6}$ $\quad 1\frac{2}{6}<1\frac{5}{6}$

Addition of Mixed Numbers

$1\frac{1}{8}+2\frac{3}{8}=3\frac{4}{8}=3\frac{1}{2}$ $\quad 1\frac{1}{3}+2\frac{3}{9}=1\frac{1}{3}+2\frac{1}{3}=3\frac{2}{3}$

$1\frac{2}{6}+2\frac{1}{9}=1\frac{1}{3}+2\frac{1}{9}=1\frac{3}{9}+2\frac{1}{9}=3\frac{4}{9}$ $\quad 1\frac{1}{2}+1\frac{1}{3}+2\frac{1}{6}=1\frac{3}{6}+1\frac{2}{6}+2\frac{1}{6}=4\frac{6}{6}=5$

Subtraction of Mixed Numbers

$6\frac{7}{8}-5\frac{3}{8}=1\frac{4}{8}=1\frac{1}{2}$ $\quad 4\frac{3}{4}-2\frac{1}{4}=2\frac{2}{4}=2\frac{1}{2}$ $\quad 3\frac{1}{2}-2=1\frac{1}{2}$

$1-\frac{2}{3}=\frac{3}{3}-\frac{2}{3}=\frac{1}{3}$ $\quad 4-\frac{5}{8}=3\frac{8}{8}-\frac{5}{8}=3\frac{3}{8}$ $\quad 8-3\frac{5}{6}=7\frac{6}{6}-3\frac{5}{6}=4\frac{1}{6}$

$6\frac{2}{7}-2\frac{5}{7}=5\frac{9}{7}-2\frac{5}{7}=3\frac{4}{7}$ $\quad 5\frac{1}{3}-2\frac{5}{6}=5\frac{2}{6}-2\frac{5}{6}=4\frac{8}{6}-2\frac{5}{6}=2\frac{3}{6}=2\frac{1}{2}$

Multiplication of Mixed Numbers

$3\frac{1}{3}\cdot 2\frac{2}{5}=\frac{10}{3}\cdot\frac{12}{5}=\frac{2\cdot 4}{1\cdot 1}=8$ $\quad \frac{5}{8}\cdot 4\frac{4}{5}\cdot\frac{1}{6}=\frac{5}{8}\cdot\frac{24}{5}\cdot\frac{1}{6}=\frac{3}{6}=\frac{1}{2}$

Division of Mixed Numbers

$1\frac{1}{6}\div 4\frac{2}{3}=\frac{7}{6}\div\frac{14}{3}=\frac{7}{6}\times\frac{3}{14}=\frac{1}{4}$ $\quad 3\frac{1}{5}\div 2\frac{2}{5}=\frac{16}{5}\times\frac{5}{12}=\frac{4}{3}=1\frac{1}{3}$

$8\div 4\frac{4}{5}=\frac{8}{1}\div\frac{24}{5}=\frac{8}{1}\times\frac{5}{24}=\frac{5}{3}=1\frac{2}{3}$ $\quad 4\frac{1}{3}\div 13=\frac{13}{3}\div\frac{13}{1}=\frac{13}{3}\times\frac{1}{13}=\frac{1}{3}$

Power of Mixed Numbers

$\left(\frac{3}{4}\right)^3=\frac{3^3}{4^3}=\frac{27}{64}$ $\quad \left(1\frac{1}{2}\right)^2=\left(\frac{3}{2}\right)^2=\frac{3^2}{2^2}=\frac{9}{4}=2\frac{1}{4}$

Grade 7 Chapter 4 Lesson 6

Mental Math 7
Fractions

1) $\frac{1}{2}\times\frac{2}{5}$

2) $\frac{1}{3}\times\frac{3}{4}$

3) $\frac{5}{8}\times 8$

4) $4\times\frac{1}{4}$

5) $\frac{1}{2}\div\frac{1}{2}$

6) $\frac{1}{3}\div\frac{1}{6}$

7) $\frac{1}{4}\div\frac{1}{12}$

8) $2\div\frac{1}{2}$

9) $(\frac{1}{2})^2$

10) $(\frac{1}{5})^2$

11) $(\frac{2}{5})^2$

12) $(\frac{3}{5})^2$

13) $\frac{1}{4}+\frac{3}{4}$

14) $\frac{1}{2}+\frac{3}{2}$

15) $\frac{7}{8}-\frac{5}{8}$

16) $\frac{7}{8}-\frac{3}{8}$

17) $\frac{1}{2}+\frac{1}{4}$

18) $\frac{3}{4}+\frac{1}{2}$

19) $\frac{1}{2}-\frac{1}{4}$

20) $\frac{3}{4}-\frac{1}{2}$

Grade 7 Chapter 4 Lesson 6

Mental Math 8
Mixed numbers

1) $1+\frac{1}{2}$

2) $\frac{1}{4}+2$

3) $1-\frac{1}{2}$

4) $2\frac{1}{2}-1$

5) $2\frac{3}{8}+1\frac{1}{8}$

6) $3\frac{3}{4}+2\frac{1}{4}$

7) $4\frac{3}{4}-2\frac{1}{4}$

8) $3\frac{5}{8}-2\frac{1}{8}$

9) $3\frac{1}{2}+2\frac{1}{2}$

10) $4\frac{1}{2}+2\frac{1}{2}$

11) $2\frac{3}{4}-1\frac{1}{4}$

12) $4\frac{5}{6}-1\frac{2}{6}$

13) $2\frac{3}{8}-1\frac{1}{8}$

14) $3\frac{3}{8}-2\frac{1}{8}$

15) $4\frac{7}{8}-3\frac{3}{8}$

16) $5\frac{5}{8}-4\frac{1}{8}$

17) $2-1\frac{1}{2}$

18) $2-1\frac{3}{4}$

19) $3-1\frac{1}{4}$

20) $3-1\frac{1}{5}$

Grade 7 Chapter 4 Lesson 7
Order of Operations Fractions and Mixed Numbers

An expression has five operations:
first, work on operations at the third level (a^n)
then, work on operations at second level (× or÷)
last, work on operations at the first level (+, -),
in order, as they occur, from left to right.

In short

Grouping Symbols	**(1) [2] {3}**
Exponents or Power – third level of operations	**a^n**
Multiplication or Division from left to right–second level	
× or ÷	**L → R**
Addition or Subtraction from left to right–first level	
+ or –	**L → R**

Examples

a) $9-2\frac{2}{5}\div\frac{8}{25}=9-\frac{12}{5}\div\frac{8}{25}=9-\frac{12}{5}\times\frac{25}{8}=9-\frac{3}{1}\times\frac{5}{2}=9-\frac{15}{2}=8\frac{2}{2}-7\frac{1}{2}=1\frac{1}{2}$

b) $\frac{14}{15}\div\left(2\frac{1}{3}-1\frac{1}{6}\right)=\frac{14}{15}\div\left(2\frac{2}{6}-1\frac{1}{6}\right)=\frac{14}{15}\div1\frac{1}{6}=\frac{14}{15}\div\frac{7}{6}=\frac{14}{15}\times\frac{6}{7}=\frac{2}{5}\times\frac{2}{1}=\frac{4}{5}$

c) $\frac{14}{17}\div\frac{7}{17}-1\frac{3}{5}=\frac{14}{17}\times\frac{17}{7}-1\frac{3}{5}=\frac{2}{1}\times\frac{1}{1}-1\frac{3}{5}=2-1\frac{3}{5}=1\frac{5}{5}-1\frac{3}{5}=\frac{2}{5}$

d) $\left(2\frac{1}{4}-1\frac{1}{2}\right)\div\frac{15}{16}=\left(1\frac{5}{4}-1\frac{2}{4}\right)\div\frac{15}{16}=\frac{3}{4}\times\frac{16}{15}=\frac{1}{1}\times\frac{4}{5}=\frac{4}{5}$

Composite Fractions

e) $\frac{\left(2+\frac{1}{3}\right)}{\left(2-\frac{5}{6}\right)}=\frac{2\frac{1}{3}}{\left(1\frac{6}{6}-\frac{5}{6}\right)}=\frac{2\frac{1}{3}}{1\frac{1}{6}}=2\frac{1}{3}\div1\frac{1}{6}=\frac{7}{3}\div\frac{7}{6}=\frac{7}{3}\times\frac{6}{7}=\frac{1}{1}\times\frac{2}{1}=\frac{2}{1}=2$ or

d) $\frac{\left(2+\frac{1}{3}\right)}{\left(2-\frac{5}{6}\right)}=\frac{\left(2+\frac{1}{3}\right)\times6}{\left(2-\frac{5}{6}\right)\times6}=\frac{2\times6+\frac{1}{3}\times\frac{6}{1}}{2\times6-\frac{5}{6}\times\frac{6}{1}}=\frac{12+2}{12-5}=\frac{14}{7}=2$

LCD(3, 6) = 6

Grade 7 Chapter 4 Lesson 7

Examples

1) $1\frac{3}{5}\times2\frac{1}{8}+1\frac{3}{5}\times2\frac{7}{8}=1\frac{3}{5}(2\frac{1}{8}+2\frac{7}{8})=\frac{8}{5}\times5=\frac{8}{5}\times\frac{5}{1}=8$

2) $4\frac{2}{3}\times\frac{3}{4}-3\frac{1}{3}\times\frac{3}{4}=\frac{3}{4}(4\frac{2}{3}-3\frac{1}{3})=\frac{3}{4}\times1\frac{1}{3}=\frac{3}{4}\times\frac{4}{3}=1$

3) $4\frac{3}{5}\times\frac{5}{6}-\frac{5}{6}\times1\frac{3}{5}=\frac{5}{6}(4\frac{3}{5}-1\frac{3}{5})=\frac{5}{6}\times3=\frac{5}{6}\times\frac{3}{1}=\frac{5}{2}=2\frac{1}{2}$

4) $4\frac{1}{2}\times\frac{1}{6}+\frac{1}{6}\times1\frac{1}{2}=\frac{1}{6}(4\frac{1}{2}+1\frac{1}{2})=\frac{1}{6}\times6=1$

5) $3\frac{3}{4}\times1\frac{1}{6}-2\frac{1}{4}\times1\frac{1}{6}=1\frac{1}{6}(3\frac{3}{4}-2\frac{1}{4})=1\frac{1}{6}\times1\frac{2}{4}=\frac{7}{6}\times\frac{3}{2}=\frac{7}{4}=1\frac{3}{4}$

6) $(\frac{1}{3}-\frac{1}{5})\div\frac{1}{15}=(\frac{1}{3}-\frac{1}{5})\times15=\frac{1}{3}\times\frac{15}{1}-\frac{1}{5}\times\frac{15}{1}=5-3=2$

7) $(1\frac{3}{4}-1\frac{1}{2})\div\frac{1}{8}=(\frac{7}{4}-\frac{3}{2})\times8=\frac{7}{4}\times\frac{8}{1}-\frac{3}{2}\times\frac{8}{1}=14-12=2$

8) $(1\frac{1}{2}-1\frac{1}{6})\div\frac{1}{6}=(\frac{3}{2}-\frac{7}{6})\times6=\frac{3}{2}\times\frac{6}{1}-\frac{7}{6}\times\frac{6}{1}=9-7=2$

9) $(2\frac{1}{2}+1\frac{1}{3})\div\frac{1}{6}=(\frac{5}{2}+\frac{4}{3})\times6=\frac{5}{2}\times\frac{6}{1}+\frac{4}{3}\times\frac{6}{1}=15+8=23$

10) $(1\frac{1}{3}+1\frac{3}{4})\div\frac{1}{12}=(\frac{4}{3}+\frac{7}{4})\times12=\frac{4}{3}\times\frac{12}{1}+\frac{7}{4}\times\frac{12}{1}=16+21=37$

Grade 7 Chapter 4 Lesson 7

Review 1 A

Perform the operations

1) $7 - 1\frac{1}{3} \div \frac{4}{15}$

2) $6 - 2\frac{1}{10} \div \frac{7}{15}$

3) $4 - 2\frac{1}{2} \div 1\frac{2}{3}$

4) $3 - 1\frac{1}{9} \div \frac{5}{12}$

5) $4\frac{3}{4} - 1\frac{2}{5} \div \frac{2}{5}$

6) $3\frac{3}{11} \div \frac{9}{11} - 2\frac{3}{8}$

7) $4\frac{1}{12} \div \frac{7}{12} - 5\frac{2}{9}$

8) $10\frac{1}{8} \div 2\frac{1}{4} - 2\frac{1}{3}$

9) $12 \div 3\frac{3}{5} - 1\frac{2}{3}$

10) $6\frac{2}{5} \div \frac{12}{35} - 16$

11) $(1\frac{5}{8} + 1\frac{1}{12}) \div 1\frac{1}{12}$

12) $(1\frac{1}{6} + 3\frac{3}{4}) \div 3\frac{5}{18}$

13) $(2\frac{1}{2} + 1\frac{1}{3}) \div \frac{1}{6}$

14) $(2\frac{2}{3} - 2\frac{2}{9}) \div 2$

15) $(1\frac{1}{2} - \frac{1}{4}) \div 3\frac{3}{4}$

16) $2\frac{7}{8} \div \left(\frac{3}{4} + 1\frac{1}{6}\right)$

17) $2\frac{1}{3} \div (4\frac{1}{6} - 3\frac{7}{9})$

18) $5\frac{3}{8} \div (5\frac{5}{16} - 2\frac{5}{8})$

19) $3\frac{1}{8} \div (3\frac{4}{6} - 1\frac{1}{6})$

20) $1\frac{8}{9} \div (3\frac{1}{6} - 1\frac{3}{4})$

Grade 7 Chapter 4 Lesson 7
Review 1 B

Solve all problems in fractions, but just final answers put in decimals

1) $8-\frac{4}{5}\div 1\frac{1}{15}$

2) $\frac{14}{15}\div\frac{7}{15}-1\frac{2}{5}$

3) $\left(3\frac{1}{4}-2\frac{1}{2}\right)\div 1\frac{1}{2}$

4) $\frac{14}{15}\div\left(3\frac{1}{3}-2\frac{1}{6}\right)$

5) $\left(1+\frac{1}{2}\right)\div\left(1-\frac{1}{2}\right)$

6) $\left(1-\frac{1}{4}\right)\div\left(1+\frac{1}{4}\right)$

7) $\dfrac{2-\frac{1}{2}}{2+\frac{1}{2}}$

8) $\dfrac{2+\frac{3}{4}}{2-\frac{3}{4}}$

9) $\dfrac{3-\frac{1}{3}}{3+\frac{1}{3}}$

10) $\dfrac{7+\frac{2}{6}}{6-\frac{2}{3}}$

11) $\dfrac{2+\frac{1}{4}}{2-\frac{1}{2}}$

12) $\dfrac{2-\frac{5}{6}}{2+\frac{1}{3}}$

13) $\dfrac{1-\frac{1}{4}}{1+\frac{7}{8}}$

14) $\dfrac{4-\frac{1}{4}}{3+\frac{1}{8}}$

15) $\dfrac{2+\frac{1}{3}}{2-\frac{5}{6}}$

16) $\dfrac{2-\frac{1}{3}}{2+\frac{1}{3}}$

17) $\dfrac{3+\frac{1}{3}}{3-\frac{1}{3}}$

18) $\dfrac{1+\frac{7}{8}}{1-\frac{1}{4}}$

19) $\dfrac{3-\frac{1}{2}}{6+\frac{1}{4}}$

20) $\dfrac{6+\frac{1}{4}}{3-\frac{1}{2}}$

Grade 7 Chapter 4 Lesson 8
Laws for Fractions and Mixed Numbers
First Law of Addition (commutative)

Changing the order of addends does not change the sum.
Add in any order:
$A + B = B + A$ $2 + 3 = 5$ $3 + 2 = 5$ $2 + 3 = 3 + 2$

Second Law of Addition (associative)

Changing the grouping of addends does not change the sum:
$A + (B + C) = (A + B) + C$ $8 + (2 + 5) = (8 + 2) + 5 = 10 + 5 = 15$

First Law of Multiplication (commutative)

Changing the order of factors does not change the product.
Multiply in any order:
$A\ B = B\ A$ $2 \times 3 = 6$ $3 \times 2 = 6$ $2 \times 3 = 3 \times 2$

Second Law of Multiplication (associative)

Changing the grouping of factors does not change the product:
$A\ (B\ C) = (AB)\ C$ $5 \times (2 \times 4) = (5 \times 2) \times 4 = 10 \times 4 = 40$

Test for Sum

If each addend can divide by number the sum divisible by a number
$(A + B) \div C = A \div C + B \div C$ $(18+12) \div 3 = 18 \div 3 + 12 \div 3 = 6 + 4 = 10$

Test for Difference

If the minuend and subtrahend can divide by number the difference divisible by a number $(A - B) \div C = A \div C - B \div C$
$(18 - 12) \div 3 = 18 \div 3 - 12 \div 3 = 6 - 4 = 2$

Distributive Laws of Multiplication

To multiply a number by the sum or difference, multiply the number by each number in parentheses, then add or subtract the products:
$A\ (B + C) = AB + AC$ $A\ (B - C) = AB - AC$
$(B + C)A = AB + AC$ $(B - C)A = AB - AC$
$14 \cdot 5 = (10 + 4) \cdot 5 = 10 \cdot 5 + 4 \cdot 5 = 50 + 20 = 70$
$18 \cdot 5 = (20 - 2) \cdot 5 = 20 \cdot 5 - 2 \cdot 5 = 100 - 10 = 90$

Common Factor (Put A before parentheses)

$AB + AC = A\ (B + C)$ $AB - AC = A\ (B - C)$
If two products have a common factor, put the common factor before parentheses. $AB + AC$ A – common factor
In parentheses, put results of divisions products by common factor
$AB \div A = B$ $AC \div A = C$
$16 \cdot 5 + 4 \cdot 5 = 5\ (16 + 4) = 5\ (20) = 100$
$22 \cdot 5 - 2 \cdot 5 = 5\ (22 - 2) = 5\ (20) = 100$

Grade 7 Chapter 4 Lesson 8
Worksheet 1 A for Final Exam

Laws of Fractions and Mixed Numbers

Evaluate by using laws

1) $1\frac{3}{5}\times 2\frac{1}{2}+1\frac{3}{5}\times 7\frac{1}{2}$

2) $4\frac{2}{3}\times\frac{3}{4}-3\frac{1}{3}\times\frac{3}{4}$

3) $4\frac{3}{5}\times\frac{5}{6}-\frac{5}{6}\times 1\frac{3}{5}$

4) $4\frac{1}{2}\times\frac{1}{6}+\frac{1}{6}\times 1\frac{1}{2}$

5) $3\frac{3}{4}\times 1\frac{1}{6}-2\frac{1}{4}\times 1\frac{1}{6}$

6) $\frac{3}{5}\times 15-\frac{1}{5}\times 15$

7) $1\frac{3}{4}\times 8-1\frac{1}{2}\times 8$

8) $1\frac{1}{2}\times 6-1\frac{1}{6}\times 6$

9) $2\frac{1}{2}\times 6-1\frac{1}{2}\times 6$

10) $1\frac{3}{4}\times 12-1\frac{1}{4}\times 12$

Worksheet 1 B for Final Exam

1) $(1\frac{3}{4}-1\frac{1}{2})\div\frac{1}{8}$

2) $(1\frac{3}{5}+1\frac{2}{5})\div\frac{1}{2}$

3) $(1\frac{3}{5}-1\frac{2}{5})\div\frac{2}{5}$

4) $(4\frac{3}{5}+1\frac{2}{5})\div\frac{1}{5}$

5) $(2\frac{1}{2}+1\frac{1}{2})\div\frac{1}{6}$

6) $(3\frac{2}{3}-1\frac{2}{3})\div\frac{2}{3}$

7) $(1\frac{4}{5}+1\frac{1}{5})\div\frac{1}{3}$

8) $(4\frac{3}{5}-1\frac{3}{5})\div\frac{1}{5}$

9) $(1\frac{1}{3}+1\frac{2}{3})\div\frac{1}{4}$

10) $(4\frac{1}{5}-2\frac{1}{5})\div\frac{1}{3}$

Grade 7 Chapter 4 Lesson 8

Review 2 A

Evaluate by using laws

1) $\frac{7}{18}+\frac{1}{3}+\frac{5}{18}$

2) $\frac{3}{14}+\frac{1}{2}+\frac{4}{14}$

3) $\frac{3}{16}+\frac{3}{4}+\frac{1}{16}$

4) $\frac{1}{15}+\frac{4}{5}+\frac{2}{15}$

5) $(\frac{1}{4}+\frac{9}{16})+\frac{7}{16}$

6) $\frac{7}{8}+\frac{3}{4}+\frac{1}{8}$

7) $\frac{3}{7}+\frac{5}{9}+\frac{4}{7}+\frac{4}{9}$

8) $(\frac{1}{7}+\frac{11}{18})+(\frac{6}{7}+\frac{7}{18})$

9) $\frac{5}{8}+(\frac{7}{24}+\frac{3}{8})+\frac{5}{24}$

10) $(\frac{8}{15}+\frac{1}{5})+(\frac{4}{15}+\frac{4}{5})$

11) $9(1-\frac{4}{9})$

12) $(\frac{2}{3}-\frac{3}{5})\times 15$

13) $9(3-\frac{5}{9})$

14) $(1+\frac{5}{6})\times 6$

15) $7(2+\frac{3}{7})$

16) $(1-\frac{1}{6})\times 6$

17) $6(1\frac{1}{2}-\frac{2}{3})$

18) $(\frac{5}{7}+\frac{4}{7})\times 7$

19) $7(\frac{5}{7}-\frac{4}{7})$

20) $(3+\frac{2}{3})\times 3$

Grade 7 Chapter 4 Lesson 8
Review 2 B
Evaluate by using laws

1) $3(2+\frac{1}{6})$

2) $(2\frac{1}{4}-\frac{3}{4})\times 4$

3) $3(1\frac{5}{6}-1\frac{1}{3})$

4) $(3\frac{1}{2}-\frac{1}{4})\times 4$

5) $3(1\frac{2}{3}-\frac{5}{6})$

6) $(1\frac{2}{7}-\frac{1}{3})\times 21$

7) $6(2\frac{1}{2}+1\frac{1}{3})$

8) $(4+\frac{5}{8})\times 2$

9) $3(1\frac{1}{3}-\frac{5}{12})$

10) $(1\frac{1}{4}-\frac{5}{8})\times 4$

Simplify by finding common factor

11) $8\times 1\frac{7}{8}-8\times\frac{3}{8}$

12) $3\frac{2}{9}\times\frac{3}{5}+\frac{3}{5}\times 6\frac{7}{9}$

13) $7\times 1\frac{3}{14}-7\times\frac{1}{14}$

14) $4\times 2\frac{7}{8}-4\times 2\frac{1}{8}$

15) $2\frac{6}{7}\times\frac{7}{10}+2\frac{1}{7}\times\frac{7}{10}$

16) $5\frac{5}{6}\times\frac{3}{4}-1\frac{5}{6}\times\frac{3}{4}$

17) $4\frac{4}{5}\times\frac{5}{8}+\frac{5}{8}\times 3\frac{1}{5}$

18) $3\frac{2}{3}\times 1\frac{1}{2}-1\frac{2}{3}\times 1\frac{1}{2}$

19) $\frac{5}{6}\times 5\frac{1}{5}-\frac{5}{6}\times 2\frac{1}{5}$

20) $3\frac{1}{3}\times 1\frac{1}{5}+1\frac{1}{5}\times 1\frac{2}{3}$

Grade 7 Chapter 4 Lesson 9
Reviews for Test 4
Review 4 A

Read directions carefully. Answers must be in the indicated place.
Show all steps for full credits. Do only 1 operation per 1 step.
Solve 1 problem 5 minutes. Each problem is worth 5 points.

Evaluate

1) Reduce fraction

$$\frac{60}{100}$$

1) ______

2) Add mixed numbers

$$2\frac{5}{12}+1\frac{5}{6}$$

2) ______

3) Subtract mixed numbers

$$4-2\frac{3}{4}$$

3) ______

4) Multiply mixed numbers

$$2\frac{2}{5}\times 5\frac{5}{6}$$

4) ______

5) Divide mixed numbers

$$4\frac{5}{8}\div 6\frac{1}{6}$$

5) ______

Grade 7 Chapter 4 Lesson 9
Review 4 A

Read directions carefully. **Answers must be in the indicated place.**
Show all steps for full credits. **Do only 1 operation per 1 step.**
Solve 1 problem 5 minutes. **Each problem is worth 5 points.**
Evaluate

6) Divide numbers

$\frac{2}{3} \div 1\frac{1}{3} \div \frac{5}{8}$

6) _____

7) Evaluate

$\left(1\frac{1}{2}\right)^3 \div \left(1\frac{1}{2}\right)$

7) _____

8) Evaluate

$\left(\frac{1}{2} - \frac{1}{3}\right)^2 \div \frac{1}{9}$

8)______

9) Evaluate

$6\frac{1}{2} \div \left(6\frac{1}{3} - 4\frac{1}{6}\right)$

9) _____

10) Evaluate

$1\frac{3}{4} \times 8 - 1\frac{1}{2} \times 8$

10)_____

Grade 7 Chapter 4 Lesson 9
Review 4 B

Read directions carefully. Answers must be in the indicated place.
Show all steps for full credits. Do only 1 operation per 1 step.
Solve 1 problem 5 minutes. Each problem is worth 5 points.
Evaluate

1) Reduce fraction

$\frac{40}{100}$

1) ______

2) Add mixed numbers

$1\frac{1}{6}+2\frac{1}{12}$

2)______

3) Subtract mixed numbers

$4-1\frac{1}{5}$

3)______

4) Multiply mixed numbers

$1\frac{5}{6}\times 2\frac{2}{11}$

4)______

5) Divide mixed numbers

$3\frac{3}{8}\div 5\frac{5}{8}$

5)______

Grade 7 Chapter 4 Lesson 9
Review 4 B

Read directions carefully. Answers must be in the indicated place.
Show all steps for full credits. Do only 1 operation per 1 step.
Solve 1 problem 5 minutes. Each problem is worth 5 points.
Evaluate

6) Divide numbers

$\frac{1}{6} \div 1\frac{1}{3} \div \frac{5}{8}$

6)______

7) Evaluate

$(1\frac{1}{2})^2 \div \left(1\frac{1}{4}\right)$

7) ______

8) Evaluate

$\left(\frac{1}{3} - \frac{1}{6}\right)^2 \div \frac{1}{18}$

8) ______

9) Evaluate

$4\frac{1}{2} \div (6\frac{1}{2} - 4\frac{1}{4})$

9) ______

10) Evaluate

$1\frac{1}{2} \times 6 - 1\frac{1}{6} \times 6$

10) ______

Grade 7 Chapter 4 Lesson 10
Tests for Chapter 4
Test 4 A

Read directions carefully. **Answers must be in the indicated place.**
Show all steps for full credits. **Do only 1 operation per 1 step.**
Solve 1 problem 5 minutes. **Each problem is worth 5 points.**

Evaluate

1) Reduce fraction

$$\frac{50}{100}$$

1)______

2) Add mixed numbers

$$1\frac{6}{15} + 2\frac{1}{5}$$

2)______

3) Subtract mixed numbers

$$5 - 1\frac{1}{5}$$

3) ______

4) Multiply mixed numbers

$$3\frac{3}{4} \times 5\frac{1}{3}$$

4)______

5) Divide mixed numbers

$$2\frac{7}{8} \div 3\frac{5}{6}$$

5) ______

Grade 7 Chapter 4 Lesson 10
Test 4 A

Read directions carefully. Answers must be in the indicated place.
Show all steps for full credits. Do only 1 operation per 1 step.
Solve 1 problem 5 minutes. Each problem is worth 5 points.
Evaluate

6) Divide numbers

$1\frac{1}{4} \div \frac{3}{4} \div 1\frac{2}{3}$

6) ________

7) Evaluate

$\left(1\frac{1}{2}\right)^3 \div \left(1\frac{1}{2}\right)^2$

7)______

8) Evaluate

$\left(\frac{1}{4} - \frac{1}{8}\right)^2 \div \frac{1}{32}$

8) _____

9) Evaluate

$2\frac{1}{2} \div \left(3\frac{1}{2} - 2\frac{1}{4}\right)$

9)______

10) Evaluate

$2\frac{1}{2} \times 16 - 1\frac{1}{4} \times 16$

10) ______

Grade 7 Chapter 4 Lesson 10
Test 4 B

Read directions carefully. **Answers must be in the indicated place.**
Show all steps for full credits. **Do only 1 operation per 1 step.**
Solve 1 problem 5 minutes. **Each problem is worth 5 points.**
Evaluate

1) Reduce fraction

$$\frac{20}{100}$$

1) ______

2) Add mixed numbers

$$1\frac{6}{15} + 1\frac{2}{5}$$

2) ______

3) Subtract mixed numbers

$$3 - 1\frac{1}{4}$$

3) ______

4) Multiply mixed numbers

$$4\frac{2}{5} \times 1\frac{4}{11}$$

4) ______

5) Divide mixed numbers

$$3\frac{3}{4} \div 6\frac{1}{4}$$

5) ______

Grade 7 Chapter 4 Lesson 10
Test 4 B

Read directions carefully. **Answers must be in the indicated place.**
Show all steps for full credits. **Do only 1 operation per 1 step.**
Solve 1 problem 5 minutes. **Each problem is worth 5 points.**

Evaluate

6) Divide numbers

$\frac{3}{8} \div \frac{3}{4} \div 1\frac{1}{4}$

6) ________

7) Evaluate

$(1\frac{4}{5}) \div \left(1\frac{1}{2}\right)^2$

7) ________

8) Evaluate

$\left(\frac{1}{5} - \frac{1}{10}\right)^2 \div \frac{1}{10}$

8) ________

9) Evaluate

$1\frac{1}{2} \div (5\frac{1}{2} - 4\frac{1}{4})$

9) ________

10) Evaluate

$1\frac{3}{4} \times 12 - 1\frac{1}{4} \times 12$

10) ________

Grade 7 Answers of Chapter 4

Lesson 1 1) $\frac{2}{5}$ 2) $\frac{4}{5}$ 3) $\frac{1}{10}$ 4) $1\frac{1}{5}$ 5) 6

Lesson 3 Review 1 A

1) $\frac{5}{6}$	2) $\frac{7}{8}$	3) 1	4) $\frac{9}{8}$	5) $<$
6) $<$	7) $=$	8) $>$	9) $2\frac{3}{4}$	10) $3\frac{3}{4}$
11) $4\frac{4}{5}$	12) $5\frac{6}{7}$	13) $\frac{11}{4}$	14) $\frac{23}{5}$	15) $\frac{35}{6}$
16) $\frac{49}{8}$	17) $3\frac{1}{4}$	18) $4\frac{2}{5}$	19) $5\frac{5}{6}$	20) $6\frac{5}{8}$
21) $\frac{10}{12}$	22) $\frac{18}{21}$	23) $\frac{28}{32}$	24) $\frac{40}{45}$	25) $\frac{2}{5}$
26) $\frac{3}{4}$	27) $\frac{3}{5}$	28) $\frac{4}{5}$	29) $\frac{3}{4}$	30) $\frac{2}{3}$
31) $\frac{5}{6}$	32) $\frac{2}{3}$	33) $\frac{2}{5}$	34) $\frac{1}{2}$	35) $\frac{5}{9}$
36) $\frac{3}{10}$	37) $\frac{3}{4}$	38) $\frac{3}{5}$	39) $\frac{4}{5}$	40) $\frac{3}{4}$

Lesson 3 Review 1 B

1) $6 = 6$	2) $6 = 6$	3) $12 = 12$	4) $24 = 24$	5) $3 < 6$
6) $8 > 4$	7) $14 < 28$	8) $40 > 24$	9) $3 < 4$	10) $10 > 6$
11) $12 < 21$	12) $40 > 30$	13) $4 < 6$	14) $8 > 6$	15) $9 < 12$
16) $12 > 8$	17) $12 < 15$	18) $28 > 24$	19) $40 < 45$	20) $66 > 60$
21) $6 > 5$	22) $3 < 14$	23) $9 > 4$	24) $7 < 12$	25) $18 < 20$
26) $40 > 18$	27) $45 < 48$	28) $84 > 40$	29) $12 = 12$	30) $10 > 6$
31) $12 < 15$	32) $24 = 24$	33) $24 > 16$	34) $40 > 20$	35) $8 = 8$
36) $9 = 9$	37) $21 < 24$	38) $42 > 39$	39) $6 = 6$	40) $22 > 20$

Grade 7 Answers of Chapter 4

Lesson 4 Review 1 A

1)	1	2)	1	3)	$\frac{1}{2}$	4)	1	5)	1
6)	$\frac{1}{2}$	7)	$\frac{2}{3}$	8)	1	9)	$\frac{1}{2}$	10)	$\frac{3}{4}$
11)	$\frac{2}{3}$	12)	$\frac{1}{2}$	13)	$\frac{1}{2}$	14)	$1\frac{1}{3}$	15)	$1\frac{1}{4}$
16)	$1\frac{1}{4}$	17)	$1\frac{1}{5}$	18)	$1\frac{1}{3}$	19)	$1\frac{1}{2}$	20)	$1\frac{1}{2}$
21)	5	22)	$4\frac{1}{4}$	23)	$3\frac{1}{2}$	24)	$3\frac{1}{2}$	25)	$5\frac{1}{4}$
26)	$4\frac{1}{6}$	27)	$7\frac{1}{3}$	28)	$7\frac{1}{4}$	29)	$6\frac{3}{4}$	30)	$6\frac{1}{6}$
31)	$5\frac{1}{2}$	32)	$6\frac{1}{2}$	33)	$3\frac{5}{6}$	34)	$3\frac{5}{8}$	35)	$3\frac{5}{6}$
36)	$6\frac{5}{8}$	37)	$4\frac{9}{10}$	38)	$4\frac{1}{2}$	39)	$6\frac{1}{4}$	40)	$6\frac{1}{3}$

Lesson 4 Review 1 B

1)	$\frac{1}{3}$	2)	$\frac{1}{2}$	3)	$\frac{2}{7}$	4)	$\frac{1}{2}$	5)	1
6)	$\frac{1}{4}$	7)	$\frac{1}{3}$	8)	$\frac{1}{3}$	9)	$\frac{2}{3}$	10)	$1\frac{1}{4}$
11)	$\frac{2}{3}$	12)	$\frac{3}{4}$	13)	$\frac{3}{4}$	14)	$\frac{2}{3}$	15)	$1\frac{4}{9}$
16)	$1\frac{3}{8}$	17)	$\frac{2}{3}$	18)	$\frac{7}{8}$	19)	$\frac{9}{10}$	20)	$\frac{5}{6}$
21)	1	22)	$\frac{3}{4}$	23)	$1\frac{1}{6}$	24)	$1\frac{1}{3}$	25)	1
26)	$1\frac{1}{2}$	27)	1	28)	$1\frac{7}{12}$	29)	$1\frac{1}{8}$	30)	$1\frac{1}{6}$
31)	$\frac{1}{2}$	32)	$1\frac{1}{6}$	33)	$1\frac{1}{12}$	34)	$\frac{3}{4}$	35)	$1\frac{1}{2}$
36)	$\frac{1}{12}$	37)	$1\frac{1}{4}$	38)	$\frac{7}{18}$	39)	$1\frac{4}{9}$	40)	$1\frac{7}{12}$

Grade 7 Answers of Chapter 4

Lesson 5 Review 1 A

1) $\frac{1}{2}$	2) $\frac{3}{5}$	3) $\frac{2}{3}$	4) $\frac{1}{2}$	5) $\frac{2}{5}$
6) $\frac{1}{3}$	7) $\frac{3}{4}$	8) $\frac{3}{5}$	9) 4	10) 9
11) 12	12) 15	13) $\frac{1}{2}$	14) $1\frac{2}{3}$	15) $2\frac{1}{3}$
16) 3	17) $8\frac{1}{3}$	18) 12	19) 3	20) 4
21) $\frac{1}{4}$	22) $\frac{1}{9}$	23) $\frac{1}{16}$	24) $\frac{1}{25}$	25) $\frac{4}{9}$
26) $\frac{4}{25}$	27) $\frac{16}{25}$	28) $\frac{25}{36}$	29) $\frac{1}{4}$	30) $\frac{4}{9}$
31) $\frac{1}{8}$	32) $\frac{8}{27}$	33) $2\frac{1}{4}$	34) $1\frac{7}{9}$	35) $2\frac{7}{9}$
36) $1\frac{9}{16}$	37) $2\frac{1}{4}$	38) $1\frac{7}{9}$	39) $2\frac{7}{9}$	40) $1\frac{9}{16}$

Lesson 5 Review 1 B

1) 9	2) 4	3) 6	4) $2\frac{1}{3}$	5) $1\frac{2}{3}$
6) 2	7) 2	8) 8	9) $\frac{2}{3}$	10) $\frac{3}{4}$
11) $1\frac{1}{5}$	12) $\frac{5}{6}$	13) $\frac{1}{6}$	14) $\frac{1}{7}$	15) $\frac{1}{8}$
16) $\frac{1}{9}$	17) 10	18) 6	19) 12	20) 8
21) $\frac{1}{4}$	22) $\frac{1}{5}$	23) $\frac{1}{6}$	24) $\frac{1}{3}$	25) $1\frac{1}{2}$
26) $1\frac{2}{5}$	27) $\frac{1}{4}$	28) $\frac{1}{6}$	29) 2	30) 3
31) 3	32) 2	33) 1	34) 1	35) 1
36) 1	37) $1\frac{2}{3}$	38) $2\frac{3}{4}$	39) $1\frac{2}{3}$	40) $1\frac{1}{2}$

Grade 7 Answers of Chapter 4

Lesson 6

Mental Math 7

1)	$\frac{1}{5}$	2)	$\frac{1}{4}$	3)	5	4)	1	5)	1
6)	2	7)	3	8)	4	9)	$\frac{1}{4}$	10)	$\frac{1}{25}$
11)	$\frac{4}{25}$	12)	$\frac{9}{25}$	13)	1	14)	2	15)	$\frac{1}{4}$
16)	$\frac{1}{2}$	17)	$\frac{3}{4}$	18)	$1\frac{1}{4}$	19)	$\frac{1}{4}$	20)	$\frac{1}{4}$

Mental Math 8

1)	$1\frac{1}{2}$	2)	$2\frac{1}{4}$	3)	$\frac{1}{2}$	4)	$1\frac{1}{2}$	5)	$3\frac{1}{2}$
6)	6	7)	$2\frac{1}{2}$	8)	$1\frac{1}{2}$	9)	6	10)	7
11)	$1\frac{1}{2}$	12)	$3\frac{1}{2}$	13)	$1\frac{1}{4}$	14)	$1\frac{1}{4}$	15)	$1\frac{1}{2}$
16)	$1\frac{1}{2}$	17)	$\frac{1}{2}$	18)	$\frac{1}{4}$	19)	$1\frac{3}{4}$	20)	$1\frac{4}{5}$

Lesson 7 Review 1 A

1)	2	2)	$1\frac{1}{2}$	3)	$2\frac{1}{2}$	4)	$\frac{1}{3}$	5)	$1\frac{1}{4}$
6)	$1\frac{5}{8}$	7)	$1\frac{7}{9}$	8)	$2\frac{1}{6}$	9)	$1\frac{2}{3}$	10)	$2\frac{2}{3}$
11)	$2\frac{1}{2}$	12)	$1\frac{1}{2}$	13)	23	14)	$\frac{2}{9}$	15)	$\frac{1}{3}$
16)	$1\frac{1}{2}$	17)	6	18)	2	19)	$1\frac{1}{4}$	20)	$1\frac{1}{3}$

Grade 7 Answers of Chapter 4

Lesson 7 Review 1 B

1)	$7\frac{1}{4}$	2)	$\frac{3}{5}$	3)	$\frac{1}{2}$	4)	$\frac{4}{5}$	5)	3
6)	$\frac{3}{5}$	7)	$\frac{3}{5}$	8)	$2\frac{1}{5}$	9)	$\frac{4}{5}$	10)	$1\frac{3}{8}$
11)	$1\frac{1}{2}$	12)	$\frac{1}{2}$	13)	$\frac{2}{5}$	14)	$1\frac{1}{5}$	15)	2
16)	$\frac{5}{7}$	17)	$1\frac{1}{4}$	18)	$2\frac{1}{2}$	19)	$\frac{2}{5}$	20)	$2\frac{1}{2}$

Lesson 8 Worksheet 1 A

1)	16	2)	1	3)	$2\frac{1}{2}$	4)	1	5)	$1\frac{3}{4}$
6)	6	7)	2	8)	2	9)	6	10)	6

Lesson 8 Worksheet 1 B

1)	2	2)	6	3)	$\frac{1}{2}$	4)	30	5)	24
6)	3	7)	9	8)	15	9)	12	10)	6

Lesson 8 Review 2 A

1)	1	2)	1	3)	1	4)	1	5)	$1\frac{1}{4}$
6)	$1\frac{3}{4}$	7)	2	8)	2	9)	$1\frac{1}{2}$	10)	$1\frac{4}{5}$
11)	5	12)	1	13)	22	14)	11	15)	17
16)	5	17)	5	18)	9	19)	1	20)	11

Review 2 B

1)	$6\frac{1}{2}$	2)	6	3)	$1\frac{1}{2}$	4)	13	5)	$2\frac{1}{2}$
6)	20	7)	23	8)	$9\frac{1}{4}$	9)	$2\frac{3}{4}$	10)	$2\frac{1}{2}$
11)	12	12)	6	13)	8	14)	3	15)	$3\frac{1}{2}$
16)	3	17)	5	18)	3	19)	$2\frac{1}{2}$	20)	6

Grade 7 Answers of Chapter 4

Lesson 9

Review 4 A

1) $\frac{3}{5}$ 2) $4\frac{1}{4}$ 3) $1\frac{1}{4}$ 4) 14 5) $\frac{3}{4}$

6) $\frac{4}{5}$ 7) $2\frac{1}{4}$ 8) $\frac{1}{4}$ 9) 3 10) 2

Review 4 B

1) $\frac{2}{5}$ 2) $3\frac{1}{4}$ 3) $2\frac{4}{5}$ 4) 4 5) $\frac{3}{5}$

6) $\frac{1}{5}$ 7) $1\frac{4}{5}$ 8) $\frac{1}{2}$ 9) 2 10) 2

Lesson 10

Test 4 A

1) $\frac{1}{2}$ 2) $3\frac{3}{5}$ 3) $3\frac{4}{5}$ 4) 20 5) $\frac{3}{4}$

6) 1 7) $1\frac{1}{2}$ 8) $\frac{1}{2}$ 9) 2 10) 20

Test 4 B

1) $\frac{1}{5}$ 2) $2\frac{4}{5}$ 3) $1\frac{3}{4}$ 4) 6 5) $\frac{3}{5}$

6) $\frac{2}{5}$ 7) $\frac{4}{5}$ 8) $\frac{1}{10}$ 9) $1\frac{1}{5}$ 10) 6

Grade 7

Chapter 5 Converting Fractions and Decimals

Grade 7 Chapter 5 Contents
Converting Fractions and Decimals
Lesson 1 – 12
Basic Problems

Work with lesson 1, read rules and examples, 2) Solve the problems from contents, 3) Check answers in the end of chapter 1, 4) Work with the next lessons 2, 3, 4, 5, …, 5) Solve Test 1 A, 6) Grading Scale: A =10 points, B=9points, C=8 points, D=7 points, F=0-6 points, 7) If you solve 10 problems correctly, you get an A, 8) If have an A, you can move on to next chapter, 9)If your grade is lower than an A, work with the mistakes, then solve Test 1B, 10) When you finish all of the chapters, solve Final Exam in last chapter.

Grade 7 Chapter 5 Lesson 1
Table 1
Multiplication and Division

$A + 0 = A$ $0 + A = A$ $A - 0 = A$ $A - A = 0$

$A \times 1 = A$ $1 \times A = A$ $A \div 1 = A$ $A \div A = 1$

$A \times 0 = 0$ $0 \times A = 0$ $0 \div A = 0$ $A \div 0 = \emptyset$

$A + B = B + A$ $A \times B = B \times A$ $\emptyset$ no solution

Number 5 from the left column multiply by number 4 from first row to get answer 20.

$5 \times 4 = 20$ $20 \div 5 = 4$ $20 \div 4 = 5$

Memorize Table 1

×	1	2	3	4	5	6	7	8	9	10
1	1	2	3	4	5	6	7	8	9	10
2	2	4	6	8	10	12	14	16	18	20
3	3	6	9	12	15	18	21	24	27	30
4	4	8	12	16	20	24	28	32	36	40
5	5	10	15	20	25	30	35	40	45	50
6	6	12	18	24	30	36	42	48	54	60
7	7	14	21	28	35	42	49	56	63	70
8	8	16	24	32	40	48	56	64	72	80
9	9	18	27	36	45	54	63	72	81	90
10	10	20	30	40	50	60	70	80	90	100

Grade 7 Chapter 5 Lesson 1
Pretest Chapter 5
Solve operations in fractions or in decimals but final answers put in decimals.

1) **Divide**

$0.57 \div \frac{3}{4}$

1) _______

2) **Divide**

$6\frac{7}{8} \div 2.75$

2) _______

Evaluate

3) $3^2 \div (1\frac{1}{2})$

3) _______

4) $(3.5 - 3\frac{1}{3}) \div 1\frac{1}{3}$

4) ________

5) $6.5 \div (3\frac{1}{3} - 1\frac{1}{6})$

5) _______

Grade 7 Chapter 5 Lesson 2

Converting Fractions to Repeating Decimals

Repeating Decimal

Decimals called repeating decimals if the decimal digits after the point repeated: $0.1818\ldots \approx 0.\overline{18}$ $\quad$ $0.1666\ldots \approx 0.1\overline{6}$

Period

A digit which is repeated is called a period.

The repeated decimal $0.\overline{18}$ has a period 18.

The repeated decimal $0.1\overline{6}$ has a period 6

Simple Repeating Decimal

If repetition digits begin at once after the decimal point called simple repeating decimal. To write a fraction to simple repeating decimals, always try to reduce the fraction first, after that, divide the numerator by the denominator.

Reduced fractions can convert to repeating decimals when the denominator's prime factorization contains not just digit 2 and/or 5.

Denominator: $6 = 2 \times 3$ $\quad$ Denominator: $33 = 3 \times 11$

$$\frac{1}{3} \approx 0.333\ldots \approx 0.\overline{3} \qquad 1\frac{4}{6} = 1\frac{2}{3} \approx 1.666\ldots \approx 1.\overline{6}$$

$$2\frac{2}{11} \approx 2.1818\ldots \approx 2.\overline{18} \qquad \frac{7}{33} \approx 0.2121\ldots \approx 0.\overline{21}$$

$$\begin{array}{r} 0.33 \\ 3\overline{)1.00} \\ -0 \\ \hline 10 \\ -9 \\ \hline 10 \\ -9 \\ \hline 1 \end{array} \qquad \begin{array}{r} 0.21 \\ 33\overline{)7.00} \\ -66 \\ \hline 40 \\ -33 \\ \hline 70 \\ -66 \\ \hline 4 \end{array}$$

Grade 7 Chapter 5 Lesson 2

Converting Fractions to Composite Repeating Decimals

Repeating Decimal

Decimals called repeating decimals if the decimal digits after the point are repeated: $0.2727\ldots \approx 0.\overline{27}$ $0.8333\ldots \approx 0.8\overline{3}$

Period

A digit which is repeated is called a period.
The repeating decimal $0.\overline{27}$ has a period 27.
The repeating decimal $0.8\overline{3}$ has a period 3.
The number with period is 83; number before the period is 8.

Simple Repeating Decimal

If repetition digits begin at once after the decimal point called simple repeating decimal:

$\frac{1}{3} \approx 0.333\ldots \approx 0.\overline{3}$ $1\frac{4}{6} = 1\frac{2}{3} \approx 1.666\ldots \approx 1.\overline{6}$ $2\frac{3}{11} \approx 2.2727\ldots \approx 2.\overline{27}$

Composite Repeating Decimal

If repetition digits begin not at once after the decimal point, after some digits, called composite repeating decimal:

$3\frac{1}{6} \approx 3.1666\ldots \approx 3.1\overline{6}$ $2\frac{5}{6} \approx 2.8333\ldots \approx 2.8\overline{3}$ $1\frac{14}{30} = 1\frac{7}{15} \approx 1.466\ldots \approx 1.4\overline{6}$

To convert a fraction to composite repeating decimals, always try to reduce the fraction first, after that, divide the numerator by the denominator.

Reduced fractions can convert to repeating decimals when the denominator's prime factorization contains not just digit 2 and/or 5.

Denominator: $14 = 2 \times 7$ **Denominator:** $30 = 6 \times 5 = 2 \times 3 \times 5$

Grade 7 Chapter 5 Lesson 2

Review 1 A

Converting Fractions to Simple Repeating Decimals

1) $\frac{4}{3}$	2) $\frac{5}{3}$	3) $\frac{4}{6}$	4) $\frac{8}{6}$
5) $\frac{1}{9}$	6) $\frac{2}{9}$	7) $\frac{3}{9}$	8) $\frac{4}{9}$
9) $\frac{5}{9}$	10) $\frac{6}{9}$	11) $\frac{7}{9}$	12) $\frac{8}{9}$
13) $\frac{1}{11}$	14) $\frac{2}{11}$	15) $\frac{3}{11}$	16) $\frac{4}{11}$
17) $\frac{5}{11}$	18) $\frac{6}{11}$	19) $\frac{7}{11}$	20) $\frac{8}{11}$
21) $\frac{9}{11}$	22) $\frac{10}{11}$	23) $\frac{4}{12}$	24) $\frac{8}{12}$
25) $\frac{10}{15}$	26) $\frac{2}{18}$	27) $\frac{4}{18}$	28) $\frac{6}{18}$
29) $\frac{8}{18}$	30) $\frac{10}{18}$	31) $\frac{12}{18}$	32) $\frac{14}{18}$
33) $\frac{16}{18}$	34) $\frac{2}{22}$	35) $\frac{4}{22}$	36) $\frac{6}{22}$
37) $\frac{8}{22}$	38) $\frac{10}{22}$	39) $\frac{12}{22}$	40) $\frac{14}{22}$

Grade 7 Chapter 5 Lesson 2

Review 1 B

Converting Fractions to Composite Repeating Decimals

1) $\frac{1}{6}$	2) $\frac{5}{6}$	3) $\frac{2}{12}$	4) $\frac{10}{12}$
5) $\frac{7}{6}$	6) $\frac{11}{6}$	7) $\frac{13}{6}$	8) $\frac{17}{6}$
9) $\frac{1}{15}$	10) $\frac{2}{15}$	11) $\frac{4}{15}$	12) $\frac{7}{15}$
13) $\frac{8}{15}$	14) $\frac{11}{15}$	15) $\frac{13}{15}$	16) $\frac{14}{15}$
17) $\frac{1}{18}$	18) $\frac{3}{18}$	19) $\frac{5}{18}$	20) $\frac{7}{18}$
21) $\frac{11}{18}$	22) $\frac{13}{18}$	23) $\frac{15}{18}$	24) $\frac{17}{18}$
25) $\frac{4}{30}$	26) $\frac{5}{30}$	27) $\frac{7}{30}$	28) $\frac{8}{30}$
29) $\frac{11}{30}$	30) $\frac{13}{30}$	31) $\frac{16}{30}$	32) $\frac{17}{30}$
33) $\frac{19}{30}$	34) $\frac{22}{30}$	35) $\frac{23}{30}$	36) $\frac{25}{30}$
37) $\frac{26}{30}$	38) $\frac{28}{30}$	39) $\frac{29}{30}$	40) $\frac{44}{30}$

Grade 7 Chapter 5 Lesson 3

Converting Fractions to Terminating Decimals

Terminating Decimals

To convert a fraction to a terminating decimal, always try to reduce the fraction first then divide the numerator by the denominator. When dividing the numerator by the denominator, if the remainder equals zero, this decimal called terminating.

$\frac{3}{10} = 0.3$ $\quad \frac{9}{100} = 0.09$ $\quad 1\frac{7}{1000} = 1.007$

$\frac{1}{2} = 0.5$ $\quad \frac{1}{4} = 0.25$ $\quad \frac{3}{4} = 0.75$

$\frac{12}{24} = \frac{1}{2} = 0.5$ $\quad \frac{6}{24} = \frac{1}{4} = 0.25$ $\quad \frac{15}{20} = \frac{3}{4} = 0.75$

Fractions can change to the terminating decimals if the fraction has a factorization of the denominator containing only digits 2 and/or 5.

Fractions can be converted to terminating decimals if the reduced fraction has denominator like 2, 4, 5, 8, 10, 16, 20, 25, 32, 40, 50, 64, 80, 100 because prime factorization for these numbers content just digits 2 and/or 5:

$\underline{4 = 2^2}$ $\quad 8 = 2^3$ $\quad \underline{10 = 2 \times 5}$ $\quad 16 = 2^4$ $\quad \underline{20 = 2^2 \times 5}$ $\quad \underline{25 = 5^2}$

$32 = 2^5$ $\quad 40 = 2^3 \times 5$ $\quad \underline{50 = 2 \times 5^2}$ $\quad 64 = 2^6$ $\quad 80 = 2^4 \times 5$ $\quad \underline{100 = 2^2 \times 5^2}$

First way to change fraction to decimal is division, lets learn second way. Denominators in decimals are 10, 100, 1000. Prime factorization these numbers are $10 = 2 \times 5$ $\quad 100 = 4 \times 25 = 2 \times 2 \times 5 \times 5$

$1000 = 8 \times 125 = 2 \times 2 \times 2 \times 5 \times 5 \times 5$ $\quad \frac{4}{5} = \frac{4 \times 2}{5 \times 2} = \frac{8}{10} = 0.8$

$\frac{7}{25} = \frac{7 \times 4}{5 \times 5 \times 2 \times 2} = \frac{28}{10^2} = \frac{28}{100} = 0.28$ $\quad \frac{9}{20} = \frac{9}{2 \times 2 \times 5} = \frac{9 \times 5}{2 \times 2 \times 5 \times 5} = \frac{9 \times 5}{100} = \frac{45}{100} = 0.45$

$\frac{3}{4} = \frac{3 \times 5 \times 5}{2 \times 2 \times 5 \times 5} = \frac{75}{100} = 0.75$ $\quad \frac{1}{125} = \frac{1 \times 8}{5 \times 5 \times 5 \times 2 \times 2 \times 2} = \frac{8}{1000} = 0.008$

Converting Mixed Number to the Decimals

$4\frac{11}{50} = 4\frac{11 \times 2}{2 \times 5 \times 5 \times 2} = 4\frac{22}{100} = 4.22$ $\quad 50 = 2 \times 5 \times 5$

Grade 7 Chapter 5 Lesson 3
Converting Decimals to Fractions
Converting Terminating Decimals to Fraction

Terminating decimals always can converted to fractions: put the digits after the decimal point in the same order in the numerator, and in the denominator, put digit 1 one time and 0 as many times as there are digits after the decimal point:

$3.2 = 3\frac{2}{10} = 3\frac{1}{5}$ $\quad$ $2.45 = 2\frac{45}{100} = 2\frac{9}{20}$

Converting Simple Repeating Decimals to Fractions

a) $x = 0.3333\ldots\ (\times 10)$ $\quad 10x = 3.333\ldots$ $\quad 10x = 3 + 0.333\ldots$

$10x = 3 + x$ $\quad 10x - x = 3$ $\quad 9x = 3$ $\quad 0.\overline{3} = \frac{3}{9} = \frac{1}{3}$

b) $x = 0.1818\ldots$ $\quad 100x = 18.18\ldots$ $\quad 100x = 18 + 0.1818\ldots$

$100x = 18 + x$ $\quad 100x - x = 18$ $\quad 99x = 18$ $\quad 0.\overline{18} = \frac{18}{99} = \frac{2}{11}$

To convert simple repeating decimals to fraction, put the period in the numerator and in the denominator, write the digit 9 as many times as there are digits in the period:

$0.\overline{3} = \frac{3}{9} = \frac{1}{3}$ $\quad 0.\overline{6} = \frac{6}{9} = \frac{2}{3}$ $\quad 2.\overline{18} = 2\frac{18}{99} = 2\frac{2}{11}$ $\quad 1.\overline{27} = 1\frac{27}{99} = 1\frac{3}{11}$

Converting Composite Repeating Decimals in the Fractions

c) $x = 0.8333\ldots\ (\times 10)$ $\quad 10x = 8.333\ldots$ $\quad 10x = 8 + 0.333\ldots$

$10x = 8 + \frac{3}{9}$ $\quad 10x = 8\frac{3}{9}$ $\quad 10x = \frac{75}{9}$

$10x = \frac{83-8}{9}$ $\quad x = \frac{83-8}{90}$ $\quad 0.8\overline{3} = \frac{75}{90} = \frac{5}{6}$

To converting composite repeating decimals $0.8\overline{3}$ to fractions, put in the numerator, the difference between the number with the period (83) and the number before the period (8). Put in the denominator, the digit 9 as many times as there are digits in the period and the digit, 0 as many times as there are digits before the period. $0.1\overline{3} = \frac{13-1}{90} = \frac{12}{90} = \frac{2}{15}$

$1.1\overline{6} = 1\frac{16-1}{90} = 1\frac{15}{90} = 1\frac{1}{6}$ $\quad 1.8\overline{3} = 1\frac{83-8}{90} = 1\frac{75}{90} = 1\frac{5}{6}$

$1.4\overline{6} = 1\frac{46-4}{90} = 1\frac{42}{90} = 1\frac{7}{15}$

Grade 7 Chapter 5 Lesson 3

Review 1 A

Converting Fractions to the Terminating Decimals

1) $\frac{1}{5}$	2) $\frac{2}{5}$	3) $\frac{3}{5}$	4) $\frac{4}{5}$
5) $\frac{6}{5}$	6) $\frac{7}{5}$	7) $\frac{8}{5}$	8) $\frac{9}{5}$
9) $\frac{1}{10}$	10) $\frac{3}{10}$	11) $\frac{7}{10}$	12) $\frac{9}{10}$
13) $\frac{1}{20}$	14) $\frac{3}{20}$	15) $\frac{7}{20}$	16) $\frac{9}{20}$
17) $\frac{11}{20}$	18) $\frac{13}{20}$	19) $\frac{17}{20}$	20) $\frac{19}{20}$
21) $\frac{13}{25}$	22) $\frac{17}{25}$	23) $\frac{19}{25}$	24) $\frac{23}{25}$
25) $\frac{11}{50}$	26) $\frac{23}{50}$	27) $\frac{37}{50}$	28) $\frac{49}{50}$
29) $\frac{11}{100}$	30) $\frac{23}{100}$	31) $\frac{37}{100}$	32) $\frac{49}{100}$
33) $1\frac{22}{50}$	34) $2\frac{32}{50}$	35) $3\frac{36}{50}$	36) $4\frac{38}{50}$
37) $5\frac{42}{50}$	38) $6\frac{44}{50}$	39) $7\frac{46}{50}$	40) $8\frac{48}{50}$

Grade 7 Chapter 5 Lesson 3

Review 1 B

Converting the Terminating Decimals to Fractions

a) $1.875 = 1\frac{875}{1000} = 1\frac{875 \div 5}{1000 \div 5} = 1\frac{175}{200} = 1\frac{175 \div 5}{200 \div 5} = 1\frac{35}{40} = 1\frac{7}{8}$

1)	0.2	2)	0.4	3)	0.6	4)	0.8
5)	1.4	6)	0.25	7)	0.75	8)	1.25
9)	1.75	10)	2.25	11)	0.125	12)	0.375
13)	0.625	14)	0.875	15)	1.125	16)	0.0625
17)	0.1875	18)	0.3125	19)	0.4375	20)	0.5625

Converting the Repeating Decimals to Fractions

b) $1.\overline{6} + 0.\overline{3} = 1\frac{6}{9} + \frac{3}{9} = 1\frac{2}{3} + \frac{1}{3} = 1\frac{3}{3} = 2$

c) $0.8\overline{3} - 0.4\overline{6} = \frac{83 - 8}{90} - \frac{46 - 4}{90} = \frac{75}{90} - \frac{42}{90} = \frac{33}{90} = \frac{11}{30}$

21)	$0.\overline{1}$	22)	$0.\overline{2}$	23)	$0.\overline{3}$	24)	$0.\overline{4}$
25)	$0.\overline{5}$	26)	$0.\overline{6}$	27)	$0.\overline{7}$	28)	$0.\overline{8}$
29)	$0.\overline{18}$	30)	$0.\overline{27}$	31)	$0.\overline{36}$	32)	$0.\overline{45}$
33)	$0.\overline{54}$	34)	$0.\overline{63}$	35)	$0.\overline{72}$	36)	$0.\overline{81}$
37)	$0.2\overline{6}$	38)	$0.5\overline{3}$	39)	$0.7\overline{3}$	40)	$0.9\overline{3}$

Grade 7 Chapter 5 Lesson 4

Operations with Fractions and Decimals

Decimals always can convert to fraction:

$$1.125 = 1\frac{125}{1000} = 1\frac{125 \div 5}{1000 \div 5} = 1\frac{25}{200} = 1\frac{25 \div 5}{200 \div 5} = 1\frac{5}{40} = 1\frac{1}{8}$$

Fractions can be converted to the terminating decimal when the denominator's prime factorization contains only digits 2 and/or 5.

Fractions $\frac{1}{2} = 0.5$ $\frac{1}{4} = 0.25$ $\frac{3}{4} = 0.75$ $\frac{1}{5} = 0.2$ $\frac{1}{8} = 0.125$ $\frac{1}{10} = 0.1$

can converted to terminating decimals. These are exact numbers.

Memorize this fractions and decimals.

To do operations with fractions and decimals, use only terminating decimals because it's exact numbers.

To do operations, always check at the fraction because the fraction cannot always converted in terminating decimal.

If a fraction can convert to a terminating decimal, work in decimals.

If a fraction cannot convert to a terminating decimal, work in fractions.

a) $1.625 = 1\frac{625}{1000} = 1\frac{625 \div 5}{1000 \div 5} = 1\frac{125}{200} = 1\frac{125 \div 5}{200 \div 5} = 1\frac{25}{40} = 1\frac{5}{8}$

b) $1.\bar{8} + 0.\bar{1} = 1\frac{8}{9} + \frac{1}{9} = 1\frac{9}{9} = 2$

c) $0.9\bar{3} - 0.5\bar{3} = \frac{93 - 9}{90} - \frac{53 - 5}{90} = \frac{84}{90} - \frac{48}{90} = \frac{36}{90} = \frac{4}{10} = 0.4$

Grade 7 Chapter 5 Lesson 4

Examples:

$16\times\frac{3}{8}=16\div 8\times 3=2\times 3=6$

$\frac{3}{4}\times 24=24\times\frac{3}{4}=24\div 4\times 3=8\times 3=24$

$\frac{2}{3}\times 0.36=0.36\times\frac{2}{3}=0.36\div 3\times 2=0.12\times 2=0.24$

$3.6\div 4\frac{1}{2}=3.6\div 4.5=36\div 45=\frac{36}{45}=\frac{4}{5}=0.8$

$(0.66+\frac{31}{50})\div(-\frac{2}{10})=(\frac{66}{100}+\frac{31}{50})\div(-\frac{1}{5})=(\frac{33}{50}+\frac{31}{50})(-\frac{5}{1})=\frac{64}{50}(-\frac{5}{1})=-6.4$

$-\ 0.25\div(\frac{39}{50}-\frac{37}{50})=-\frac{1}{4}\div\frac{2}{50}=-\frac{1}{4}\times\frac{25}{1}=-\frac{25}{4}=-6\frac{1}{4}=-6.25$

$(-\frac{43}{50}+\frac{41}{50})\div 0.16=(-\frac{2}{50})\div\frac{16}{100}=-\frac{1}{25}\div\frac{4}{25}=-\frac{1}{25}\times\frac{25}{4}=-\frac{1}{4}=-0.25$

$9.6\div(-\frac{49}{50}-\frac{47}{50})=9.6\div(-\frac{96}{50})=9\frac{6}{10}\times(-\frac{50}{96})=9\frac{3}{5}(-\frac{25}{48})=-\frac{48}{5}\times\frac{25}{48}=-5$

Addition and subtraction is easy work in decimals.

Multiplication and division is easy work in fractions.

Always try to find the easiest way to calculate, that way, fewer mistakes will be made.

$\frac{1}{2}-0.25=0.5-0.25=0.25$

$0.25\div\frac{1}{8}=\frac{1}{4}\div\frac{1}{8}=\frac{1}{4}\times\frac{8}{1}=2$

Grade 7 Chapter 5 Lesson 4

Review 1 A

1) $0.75 + \frac{4}{5}$

2) $1\frac{1}{5} + 0.25$

3) $0.75 + 1\frac{2}{5}$

4) $1\frac{3}{5} + 0.375$

5) $1\frac{4}{5} + 0.625$

6) $0.875 + \frac{3}{6}$

7) $1\frac{3}{6} + 1.125$

8) $0.125 + \frac{4}{8}$

9) $\frac{6}{12} - 0.375$

10) $\frac{6}{15} - 0.25$

11) $\frac{12}{15} - 0.75$

12) $\frac{8}{16} - 0.35$

13) $\frac{3}{4} - 0.45$

14) $1\frac{1}{4} - 0.45$

15) $2\frac{1}{4} - 1.75$

16) $4\frac{1}{2} - 0.85$

17) $\frac{6}{8} \times 0.4$

18) $0.6 \times \frac{3}{12}$

19) $\frac{9}{12} \times 0.8$

20) $1.2 \times \frac{4}{16}$

Grade 7 Chapter 5 Lesson 4
Review 1 B

1) $2.4\times\frac{12}{16}$

2) $\frac{3}{8}\times1.6$

3) $3.2\times\frac{5}{8}$

4) $\frac{7}{8}\times4.8$

5) $1\frac{1}{8}\times0.8$

6) $1.6\times\frac{2}{16}$

7) $\frac{4}{16}\times3.2$

8) $6.4\times\frac{8}{16}$

9) $0.3\div\frac{6}{16}$

10) $0.5\div\frac{10}{16}$

11) $0.7\div\frac{14}{16}$

12) $0.3\div\frac{3}{5}$

13) $0.25\div\frac{4}{16}$

14) $0.15\div\frac{12}{16}$

15) $0.45\div\frac{12}{16}$

16) $0.35\div\frac{4}{16}$

17) $0.375\div\frac{10}{16}$

18) $3.75\div3\frac{1}{8}$

19) $4\frac{1}{8}\div2.75$

20) $2\frac{1}{4}\div1.875$

Grade 7 Chapter 5 Lesson 5

Order of Operations

If an expression has five operations,

first work on operations at the third level of operations (a^n),

then work on the operations at second level of operations (×,÷).

last, work on operations at first level of operations (+,-),

in order, as they occur, from left to right.

in short

Grouping Symbols **(1) [2] {3}**

Exponents or Power – third level of operations a^n

Multiplication or Division from left to right–second level

× **or** ÷ **L → R**

Addition or Subtract from left to right – first level

+ **or** – **L → R**

Always try to reduce fractions first, convert improper fractions to mixed numbers, and convert fractions to terminating decimals.

Work in fractions or decimals, but if you can the final answers should give in decimals. Try to find easiest way.

Grade 7 Chapter 5 Lesson 5

Review 1 A (#1 - # 15)

Perform the Operations

1) $(1.25 - \frac{1}{4}) \div \frac{2}{7}$

2) $(1.08 - \frac{2}{25}) \div \frac{4}{7}$

3) $(\frac{1}{6} + \frac{1}{3}) \div 0.25$

4) $(0.3 - \frac{3}{20}) \div \frac{1}{20}$

5) $(2.5 - 2\frac{1}{3}) \div \frac{5}{6}$

6) $(4.5 - 1\frac{1}{6}) \div \frac{5}{9}$

7) $(1.5 - 1\frac{1}{3}) \div \frac{1}{3}$

8) $1.2 \div (\frac{8}{9} - \frac{2}{3})$

9) $1.5 \div (\frac{5}{6} - \frac{2}{3})$

10) $3.6 \div (3\frac{1}{5} + 2\frac{4}{5})$

11) $3.5 \div (3\frac{1}{5} - 2\frac{7}{10})$

12) $2.4 \div (2\frac{3}{4} + 1\frac{1}{4})$

13) $6.75 \div (1\frac{1}{8} + 2\frac{1}{4})$

14) $29 \div (2.5 + 1\frac{1}{8})$

15) $4.2 \div (3\frac{1}{2} + 1\frac{3}{4})$

16) $1.1 \div (3\frac{1}{2} - 1\frac{3}{10})$

Review 1 B (# 16 - # 30)

17) $4.2 \div (3\frac{1}{5} - 1\frac{1}{10})$

18) $\frac{5}{6} \div (1\frac{2}{3} - 1.25)$

19) $6.5 \div (4\frac{1}{3} - 2\frac{1}{6})$

20) $\frac{8}{15} \div (1\frac{1}{6} + 1.5)$

21) $\frac{7}{15} \div (5\frac{5}{6} - 3.5)$

22) $\frac{4}{15} \div (2.8 - 2\frac{2}{3})$

23) $\frac{5}{6} \div (1\frac{8}{12} - 1.25)$

24) $\frac{8}{15} \div (1\frac{1}{6} + 1.5)$

25) $\frac{7}{15} \div (5\frac{5}{6} - 3.5)$

26) $1\frac{2}{3} \div (7.5 - 3\frac{1}{3})$

27) $2.8 \div (2\frac{3}{4} + 1.25)$

28) $2\frac{1}{9} \div (\frac{4}{9} + 0.4)$

29) $0.4 \div (\frac{8}{9} - \frac{2}{3})$

30) $3\frac{4}{5} \div (3\frac{1}{3} - 2.7)$

Grade 7 Chapter 5 Lesson 6
Laws for operations
Distributive laws of multiplication

(A + B) C = AC + BC **(A - B) C = AC – B C**

Example a) $\frac{5}{6}\times 3=\frac{5}{6}\times\frac{3}{1}=\frac{5}{2}=2\frac{1}{2}=2.5$

Example b) $\frac{5}{3}\times 6=6\times\frac{5}{3}=6\div 3\times 5=2\times 5=10$

Example c) $\left(1\frac{1}{3}-\frac{5}{6}\right)\times 3=\frac{4}{3}\times\frac{3}{1}-\frac{5}{6}\times\frac{3}{1}=4-\frac{5}{2}=4-2.5=1.5$

Review 1 A
Evaluate by using laws

1) $9(1-\frac{4}{9})$

2) $(\frac{2}{3}-\frac{3}{5})\times 15$

3) $9(3-\frac{5}{9})$

4) $(1+\frac{5}{6})\times 6$

5) $7(2+\frac{3}{7})$

6) $(1-\frac{1}{6})\times 6$

7) $6(1\frac{1}{2}-\frac{2}{3})$

8) $(\frac{5}{7}+\frac{4}{7})\times 7$

9) $7(\frac{5}{7}-\frac{4}{7})$

10) $(3+\frac{2}{3})\times 3$

11) $3(2+\frac{1}{6})$

12) $(2\frac{1}{4}-\frac{5}{8})\times 4$

13) $3(1\frac{5}{6}-1\frac{1}{3})$

14) $(3\frac{1}{2}-\frac{1}{4})\times 4$

15) $3(1\frac{2}{3}-\frac{5}{6})$

16) $(1\frac{2}{7}-\frac{1}{3})\times 21$

17) $6(2\frac{1}{2}+1\frac{1}{3})$

18) $(4+\frac{5}{8})\times 2$

19) $3(1\frac{1}{3}-\frac{5}{12})$

20) $(1\frac{1}{4}-\frac{5}{8})\times 4$

Grade 7 Chapter 5 Lesson 6

Review 1 B

Solve by finding common factor

Common Factor (Put C before parentheses)

AC + BC = C (A + B) AC – B C = C (A - B)

Example $1\frac{1}{4}\times\frac{3}{4}+\frac{3}{4}\times2\frac{3}{4}=\frac{3}{4}(1\frac{1}{4}+2\frac{3}{4})=\frac{3}{4}\times\frac{4}{1}=3$

Evaluate by using laws

1) $8\times1\frac{3}{4}-8\times\frac{5}{8}$

2) $3\frac{2}{9}\times\frac{3}{5}+\frac{3}{5}\times6\frac{7}{9}$

3) $7\times1\frac{1}{7}-7\times\frac{5}{14}$

4) $4\times2\frac{1}{4}-4\times1\frac{7}{8}$

5) $2\frac{6}{7}\times\frac{7}{10}+2\frac{1}{7}\times\frac{7}{10}$

6) $5\frac{5}{6}\times\frac{3}{4}-1\frac{5}{6}\times\frac{3}{4}$

7) $4\frac{4}{5}\times\frac{5}{8}+\frac{5}{8}\times3\frac{1}{5}$

8) $3\frac{2}{3}\times1\frac{1}{2}-1\frac{2}{3}\times1\frac{1}{2}$

9) $\frac{5}{6}\times5\frac{1}{5}-\frac{5}{6}\times2\frac{1}{5}$

10) $3\frac{1}{3}\times1\frac{1}{5}+1\frac{1}{5}\times1\frac{2}{3}$

11) $1\frac{3}{5}\times2\frac{1}{2}+1\frac{3}{5}\times7\frac{1}{2}$

12) $4\frac{2}{3}\times\frac{3}{4}-3\frac{1}{3}\times\frac{3}{4}$

13) $4\frac{3}{5}\times\frac{5}{6}-\frac{5}{6}\times1\frac{3}{5}$

14) $4\frac{1}{2}\times\frac{1}{6}+\frac{1}{6}\times1\frac{1}{2}$

15) $3\frac{3}{4}\times1\frac{1}{6}-2\frac{1}{4}\times1\frac{1}{6}$

16) $\frac{1}{3}\times15-\frac{1}{5}\times15$

17) $1\frac{3}{4}\times8-1\frac{1}{2}\times8$

18) $1\frac{1}{2}\times6-1\frac{1}{6}\times6$

19) $2\frac{1}{2}\times6-1\frac{1}{2}\times6$

20) $1\frac{3}{4}\times12-1\frac{1}{4}\times12$

21) $1\frac{3}{4}+1\frac{1}{4})\div\frac{1}{4}$

22) $(1\frac{3}{5}-1\frac{2}{5})\div\frac{2}{5}$

23) $(4\frac{3}{5}+1\frac{2}{5})\div\frac{1}{5}$

24) $(1\frac{3}{4}-1\frac{1}{4})\div\frac{1}{8}$

25) $(2\frac{1}{2}+1\frac{1}{2})\div\frac{1}{6}$

26) $(3\frac{2}{3}-1\frac{2}{3})\div\frac{2}{3}$

27) $(1\frac{5}{6}+1\frac{1}{6})\div\frac{1}{6}$

28) $(4\frac{3}{5}-1\frac{3}{5})\div\frac{1}{5}$

29) $(1\frac{2}{3}+1\frac{1}{3})\div\frac{1}{3}$

30) $(4\frac{5}{6}-2\frac{1}{6})\div\frac{1}{6}$

Grade 7 Chapter 5 Lesson 7
Order of operations and laws for numbers
Order of operations shortly

1) (1) [2] {3}

2) a^n

3) × or ÷ L → R

4) + or – L → R

If you can, always in answer try to reduce fractions to lowest term, convert improper fraction to mixed number.
If you can, do operations in fractions or decimals but final answers put in decimals. Easy work: + and – in decimals, × and ÷ in fractions.

$\frac{3}{4} + 0.25 = 0.75 + 0.25 = 1$ $\qquad 0.5 \div (\frac{1}{8}) = \frac{1}{2} \div (\frac{1}{8}) = \frac{1}{2} \times (\frac{8}{1}) = 4$

Review 1 A

1) $(1\frac{3}{8} + 1.75) \div 6\frac{1}{4}$

2) $(1.75 - 1\frac{5}{8}) \div \frac{5}{8}$

3) $(1\frac{1}{2} + 1.8) \div 2.2$

4) $(2\frac{3}{5} + 1.6) \div 2.4$

5) $1.65 \div (0.6 + \frac{1}{2})$

6) $2.2 \div (3\frac{2}{5} - 1.2)$

7) $1\frac{7}{8} \div (1\frac{1}{5} + 1.3)$

8) $2.5 \div (4.5 - 3\frac{1}{4})$

9) $5\frac{1}{4} \div (2\frac{1}{4} + 1.25)$

10) $3.3 \div (4\frac{1}{2} - 3.4)$

Grade 7 Chapter 5 Lesson 7

Laws for numbers
Common Factor

AB + AC = A (B + C)   AB - AC = A (B - C)
$16 \cdot 5 + 4 \cdot 5 = 5\,(16 + 4) = 5\,(20) = 100$
$22 \cdot 5 - 2 \cdot 5 = 5\,(22 - 2) = 5\,(20) = 100$

Review 1 B
Evaluate

1) $1\frac{3}{5}(2\frac{1}{8}) + 1\frac{3}{5}(2\frac{7}{8})$

2) $\frac{3}{4}(4\frac{2}{3}) - \frac{3}{4}(3\frac{1}{3})$

3) $\frac{5}{6}(4\frac{3}{5}) - \frac{5}{6} \times 1\frac{3}{5}$

4) $4\frac{1}{2} \times \frac{1}{6} + \frac{1}{6} \times 1\frac{1}{2}$

5) $3\frac{3}{4} \times 1\frac{1}{6} - 2\frac{1}{4} \times 1\frac{1}{6}$

6) $(\frac{1}{3} - \frac{1}{6}) \div \frac{1}{9}$

7) $(1\frac{3}{4} - 1\frac{1}{2}) \div \frac{1}{8}$

8) $(1\frac{1}{2} - 1\frac{1}{6}) \div \frac{1}{9}$

9) $(2\frac{1}{2} + 1\frac{1}{3}) \div \frac{1}{6}$

10) $(1\frac{1}{3} + 1\frac{3}{4}) \div \frac{1}{12}$

Grade 7 Chapter 5 Lesson 8

Worksheet 3 A for Final Exam

Perform the indicated operations

1) $\left(0.25+\frac{1}{8}\right)\div 1.5$

2) $\left(\frac{1}{16}+\frac{1}{32}\right)\div\frac{3}{16}$

3) $\left(0.1+\frac{1}{25}\right)\div 1.4$

4) $\left(\frac{1}{64}+\frac{1}{32}\right)\div\frac{3}{16}$

5) $\left(0.05+\frac{1}{40}\right)\div\frac{3}{20}$

6) $\left(0.02+\frac{1}{25}\right)\div\frac{6}{40}$

7) $\left(\frac{1}{80}+\frac{1}{40}\right)\div\frac{3}{64}$

8) $\left(0.01+\frac{1}{50}\right)\div\frac{1}{20}$

9) $\left(0.01+\frac{1}{25}\right)\div 0.25$

10) $\left(0.01+\frac{1}{20}\right)\div\frac{12}{20}$

Grade 7 Chapter 5 Lesson 8

Worksheet 3 B for Final Exam

Perform the indicated operations

1) $\frac{3}{8} \div \left(0.25 - \frac{1}{8}\right)$

2) $\frac{5}{32} \div \left(\frac{1}{16} - \frac{1}{32}\right)$

3) $\frac{6}{50} \div \left(0.1 - \frac{1}{25}\right)$

4) $\frac{5}{32} \div \left(\frac{1}{32} - \frac{1}{64}\right)$

5) $\frac{3}{20} \div \left(0.05 - \frac{1}{40}\right)$

6) $\frac{7}{25} \div \left(\frac{1}{25} - 0.02\right)$

7) $\frac{13}{40} \div \left(\frac{1}{40} - \frac{1}{80}\right)$

8) $\frac{9}{50} \div \left(\frac{1}{50} - 0.01\right)$

9) $\frac{9}{25} \div \left(\frac{1}{25} - 0.01\right)$

10) $\frac{8}{25} \div \left(\frac{1}{20} - 0.01\right)$

Grade 7 Chapter 5 Lesson 9
Reviews for Chapter 5
Review 5 A

Read directions carefully. Answers must be in the indicated place.
Show all steps for full credits. Do only 1 operation per 1 step.
Solve 1 problem 5 minutes. Each problem is worth 5 points.

Evaluate

1) Convert decimal 0.125 to fraction

1) ______

2) Convert fraction to terminating decimal

$\frac{3}{25}$

2) ______

3) Add $0.875 + \frac{3}{6}$

3) ______

4) Subtract $1\frac{1}{4} - 0.45$

4) ______

5) Multiply $1.6 \times \frac{3}{8}$

5) ______

Grade 7 Chapter 5 Lesson 9
Review 5 A

Read directions carefully. **Answers must be in the indicated place.**
Show all steps for full credits. **Do only 1 operation per 1 step.**
Solve 1 problem 5 minutes. **Each problem is worth 5 points.**

Evaluate

6) $0.15 \div \frac{3}{4}$

6)______

7) $3.75 \div 3\frac{1}{8}$

7)______

8) $3^3 \div \left(1\frac{1}{2}\right)^2$

8)______

9) $(0.3 - \frac{3}{20}) \div \frac{1}{20}$

9) ______

10) $\frac{5}{6} \div (1\frac{2}{3} - 1.25)$

10) ______

Grade 7 Chapter 5 Lesson 9
Review 5 B

Read directions carefully. Answers must be in the indicated place.
Show all steps for full credits. Do only 1 operation per 1 step.
Solve 1 problem 5 minutes. Each problem is worth 5 points.
Evaluate

1) Convert decimal 0.375 to fraction

1) ______

2) Convert fraction to terminating decimal

$\frac{7}{25}$

2) ______

3) Add $1\frac{3}{6}+1.125$

3) ______

4) Subtract $2\frac{1}{4}-1.75$

4) ______

5) Multiply $3.2\times\frac{5}{8}$

5) ______

Grade 7 Chapter 5 Lesson 9
Review 5 B

Read directions carefully. Answers must be in the indicated place.
Show all steps for full credits. Do only 1 operation per 1 step.
Solve 1 problem 5 minutes. Each problem is worth 5 points.

Evaluate

6) $0.45 \div \frac{3}{4}$

6) ______

7) $4\frac{1}{8} \div 2.75$

7) ______

8) $(1\frac{1}{2})^2 \div 3^2$

8) ______

9) $(1.5 - 1\frac{1}{3}) \div \frac{1}{3}$

9) ______

10) $6.5 \div (4\frac{1}{3} - 2\frac{1}{6})$

10) ______

Grade 7 Chapter 5 Lesson 10
Tests for Chapter 5
Test 5 A

Read directions carefully. Answers must be in the indicated place.
Show all steps for full credits. Do only 1 operation per 1 step.
Solve 1 problem 5 minutes. Each problem is worth 5 points.

Evaluate

1) Convert decimal 0.625 to fraction

1)______

2) Convert fraction to terminating decimal
$\frac{9}{25}$

2) ________

3) Add $1.25 + \frac{4}{8}$

3) ________

4) Subtract $1\frac{1}{4} - 0.75$

4) ________

5) Multiply $2.4 \times \frac{3}{8}$

5) ________

Grade 7 Chapter 5 Lesson 10
Test 5 A

Read directions carefully. Answers must be in the indicated place.
Show all steps for full credits. Do only 1 operation per 1 step.
Solve 1 problem 5 minutes. Each problem is worth 5 points.
Evaluate

6) $0.36 \div \frac{3}{4}$

6)______

7) $4.25 \div 2\frac{1}{8}$

7) ________

8) $3^2 \div \left(1\frac{1}{2}\right)^2$

8) _______

9) $(0.6 - \frac{3}{20}) \div \frac{3}{20}$

9) _______

10) $\frac{5}{6} \div (2\frac{2}{3} - 2.25)$

10) ______

Grade 7 Chapter 5 Lesson 10
Test 5 B

Read directions carefully. Answers must be in the indicated place.
Show all steps for full credits. Do only 1 operation per 1 step.
Solve 1 problem 5 minutes. Each problem is worth 5 points.

Evaluate

1) Convert 0.875 decimal to fraction

1) ______

2) Convert fraction to terminating decimal
$\frac{11}{25}$

2) ______

3) Add $1\frac{4}{8}+1.25$

3) ______

4) Subtract $2\frac{1}{4}-1.75$

4) ______

5) Multiply $4.8\times\frac{5}{8}$

5) ______

Grade 7 Chapter 5 Lesson 10
Test 5 B

Read directions carefully. **Answers must be in the indicated place.**
Show all steps for full credits. **Do only 1 operation per 1 step.**
Solve 1 problem 5 minutes. **Each problem is worth 5 points.**

Evaluate

6) $0.57 \div \frac{3}{4}$

6) _______

7) $6\frac{7}{8} \div 2.75$

7) _______

8) $3^2 \div (1\frac{1}{2})$

8) ______

9) $(3.5 - 3\frac{1}{3}) \div 1\frac{1}{3}$

9) ______

10) $6.5 \div (3\frac{1}{3} - 1\frac{1}{6})$

10) ______

Grade 7 Chapter 5 Lesson 11
Review 7 A
(Basic Problems) # 1 - # 10

Read directions carefully. Answers must be in the indicated place.
Show all steps for full credits. Do only 1 operation per 1 step.
Solve 1 problem 5 minutes. Each problem is worth 5 points.

Evaluate and put final answers in decimals

1) $4.3 \times 3 + 3.5 \times 3 - 5.8 \times 3$

1)______

2) $\left(3\frac{1}{2} - 2\frac{1}{4}\right) \div 1\frac{1}{4}$

2) ______

3) $\dfrac{3+\frac{1}{3}}{3-\frac{1}{3}}$

3) ______

4) $1\frac{3}{5} \times 2\frac{1}{2} + 1\frac{3}{5} \times 7\frac{1}{2}$

4) ______

5) $\left(1\frac{4}{5} + 1\frac{1}{5}\right) \div \frac{1}{3}$

5) ______

Grade 7 Chapter 5 Lesson 11
Review 7 A
(Basic Problems) # 1 - # 10

Read directions carefully. **Answers must be in the indicated place.**
Show all steps for full credits. **Do only 1 operation per 1 step.**
Solve 1 problem 5 minutes. **Each problem is worth 5 points.**

6) **Convert fraction** $\frac{1}{8}$ **to decimal**

6)______

Evaluate and put final answers in decimals

7) $6\left(2\frac{1}{2}+1\frac{1}{3}\right)$

7) ______

8) $2\frac{6}{7}\times\frac{7}{10}+2\frac{1}{7}\times\frac{7}{10}$

8) ____ _

9) $\left(1\frac{3}{4}-1\frac{1}{2}\right)\div\frac{1}{8}$

9) _______

10) $\frac{13}{40}\div\left(\frac{1}{40}-\frac{1}{80}\right)$

10) _____

Grade 7 Chapter 5 Lesson 11
Review 7 B
(Basic Problems) # 1 - # 10

Read directions carefully. **Answers must be in the indicated place.**
Show all steps for full credits. **Do only 1 operation per 1 step.**
Solve 1 problem 5 minutes. **Each problem is worth 5 points.**

Evaluate and put final answers in decimals

1) $4.9 \times 7 - 1.8 \times 7 + 6.9 \times 7$

1) ______

2) $\frac{14}{15} \div \left(3\frac{1}{3} - 2\frac{1}{6}\right)$

2) ______

3) $\dfrac{1 + \frac{7}{8}}{1 - \frac{1}{4}}$

3) ____

4) $4\frac{2}{3} \times \frac{3}{4} - 3\frac{1}{3} \times \frac{3}{4}$

4) _______

5) $\left(4\frac{3}{5} - 1\frac{3}{5}\right) \div \frac{1}{5}$

5)______

Grade 7 Chapter 5 Lesson 11
Review 7 B
(Basic Problems) # 1 - # 10

Read directions carefully. Answers must be in the indicated place.
Show all steps for full credits. Do only 1 operation per 1 step.
Solve 1 problem 5 minutes. Each problem is worth 5 points.

6) Convert fraction $\frac{3}{8}$ to decimal

6) ____

Evaluate and put final answers in decimals

7) $\left(4+\frac{5}{8}\right)\times 2$

7) ________

8) $5\frac{5}{6}\times\frac{3}{4}-1\frac{5}{6}\times\frac{3}{4}$

8) _______

9) $\left(1\frac{1}{2}-1\frac{1}{6}\right)\div\frac{1}{9}$

9) _______

10) $\frac{9}{50}\div\left(\frac{1}{50}-0.01\right)$

10) _____

Grade 7 Chapter 5 Lesson 12
Final Exam 7 A
(Basic Problems) # 1 - # 10

Read directions carefully. Answers must be in the indicated place.
Show all steps for full credits. Do only 1 operation per 1 step.
Solve 1 problem 5 minutes. Each problem is worth 5 points.

Evaluate and put final answers in decimals

1) $5.6 \times 4 + 2.7 \times 4 - 8.3 \times 4$

1)______

2) $\left(1+\frac{1}{2}\right) \div \left(1-\frac{1}{2}\right)$

2) ______

3) $\dfrac{3-\frac{1}{2}}{6+\frac{1}{4}}$

3) ______

4) $4\frac{3}{5} \times \frac{5}{6} - \frac{5}{6} \times 1\frac{3}{5}$

4) ______

5) $\left(1\frac{1}{3}+1\frac{2}{3}\right) \div \frac{1}{4}$

5) ______

Grade 7 Chapter 5 Lesson 12
Final Exam 7 A
(Basic Problems) # 1 - # 10

Read directions carefully. Answers must be in the indicated place.
Show all steps for full credits. Do only 1 operation per 1 step.
Solve 1 problem 5 minutes. Each problem is worth 5 points.

6) Convert $\frac{5}{8}$ to fraction

6)_____

Evaluate and put final answers in decimals

7) $3\left(1\frac{5}{6}-1\frac{1}{3}\right)$

7) _______

8) $4\frac{4}{5}\times\frac{5}{8}+\frac{5}{8}\times3\frac{1}{5}$

8) _______

9) $\left(2\frac{1}{2}+1\frac{1}{3}\right)\div\frac{1}{6}$

9) _______

10) $\frac{9}{25}\div\left(\frac{1}{25}-0.01\right)$

10) ______

Grade 7 Chapter 5 Lesson 12
Final Exam 7 B (Basic Problems)

Read directions carefully. Answers must be in the indicated place.
Show all steps for full credits. Do only 1 operation per 1 step.
Solve 1 problem 5 minutes. Each problem is worth 5 points.
Evaluate and put final answers in decimals

1) $3.4 \times 5 - 1.7 \times 5 + 8.3 \times 5$

1) _____

2) $\left(1 - \frac{1}{4}\right) \div \left(1 + \frac{1}{4}\right)$

2) _____

3) $\dfrac{6 + \frac{1}{4}}{3 - \frac{1}{2}}$

3) _____

4) $4\frac{1}{2} \times \frac{1}{6} + \frac{1}{6} \times 1\frac{1}{2}$

4) _____

5) $\left(4\frac{1}{5} - 2\frac{1}{5}\right) \div \frac{1}{3}$

5)______

Final Exam 7 B (Basic Problems)

Read directions carefully. Answers must be in the indicated place.
Show all steps for full credits. Do only 1 operation per 1 step.
Solve 1 problem 5 minutes. Each problem is worth 5 points.

Evaluate and put final answers in decimals

6) Convert fraction $\frac{7}{8}$ to decimal

6)____

Evaluate, put final answers in decimals

7) $\left(3\frac{1}{2}-\frac{1}{4}\right)\times 4$

7) ____

8) $3\frac{2}{3}\times 1\frac{1}{2}-1\frac{2}{3}\times 1\frac{1}{2}$

8) ______

9) $\left(1\frac{1}{3}+1\frac{3}{4}\right)\div\frac{1}{12}$

9)

10) $\frac{8}{25}\div\left(\frac{1}{20}-0.01\right)$

10) ______

Grading Scale: A =10 points, B=9points, C=8 points, D=7 points, F=0-6 points. Lesson 12 Final Exam 7 (Basic Problems) # 1 – 10 all
7 B 1) 50 2) 0.6 3)2.5 4) 1 5) 6 6) 0.875 7) 13 8)3 9)37 10) 8

Grade 7 Answers Chapter 5

Lesson 1 Pretest 1) 0.76 2) 2.5 3) 6 4) 0.125 5) 3

Lesson 2 Converting fractions to simple repeating decimals. R 1A

1)	$1.\overline{3}$	2)	$1.\overline{6}$	3)	$0.\overline{6}$	4)	$1.\overline{3}$
5)	$0.\overline{1}$	6)	$0.\overline{2}$	7)	$0.\overline{3}$	8)	$0.\overline{4}$
9)	$0.\overline{5}$	10)	$0.\overline{6}$	11)	$0.\overline{7}$	12)	$0.\overline{8}$
13)	$0.\overline{09}$	14)	$0.\overline{18}$	15)	$0.\overline{27}$	16)	$0.\overline{36}$
17)	$0.\overline{45}$	18)	$0.\overline{54}$	19)	$0.\overline{63}$	20)	$0.\overline{72}$
21)	$0.\overline{81}$	22)	$0.\overline{90}$	23)	$0.\overline{3}$	24)	$0.\overline{6}$
25)	$0.\overline{6}$	26)	$0.\overline{1}$	27)	$0.\overline{2}$	28)	$0.\overline{3}$
29)	$0.\overline{4}$	30)	$0.\overline{5}$	31)	$0.\overline{6}$	32)	$0.\overline{7}$
33)	$0.\overline{8}$	34)	$0.\overline{09}$	35)	$0.\overline{18}$	36)	$0.\overline{27}$
37)	$0.\overline{36}$	38)	$0.\overline{45}$	39)	$0.\overline{54}$	40)	$0.\overline{63}$

Lesson 2 Converting fractions to composite repeating decimals. R 1B

1)	$0.1\overline{6}$	2)	$0.8\overline{3}$	3)	$0.1\overline{6}$	4)	$0.8\overline{3}$
5)	$1.1\overline{6}$	6)	$1.8\overline{3}$	7)	$2.1\overline{6}$	8)	$2.8\overline{3}$
9)	$0.0\overline{6}$	10)	$0.1\overline{3}$	11)	$0.2\overline{6}$	12)	$0.4\overline{6}$
13)	$0.5\overline{3}$	14)	$0.7\overline{3}$	15)	$0.8\overline{6}$	16)	$0.9\overline{3}$
17)	$0.0\overline{5}$	18)	$0.1\overline{6}$	19)	$0.2\overline{7}$	20)	$0.3\overline{8}$
21)	$0.6\overline{1}$	22)	$0.7\overline{2}$	23)	$0.8\overline{3}$	24)	$0.9\overline{4}$
25)	$0.1\overline{3}$	26)	$0.1\overline{6}$	27)	$0.2\overline{3}$	28)	$0.2\overline{6}$
29)	$0.3\overline{6}$	30)	$0.4\overline{3}$	31)	$0.5\overline{3}$	32)	$0.5\overline{6}$
33)	$0.6\overline{3}$	34)	$0.7\overline{3}$	35)	$0.7\overline{6}$	36)	$0.8\overline{3}$
37)	$0.8\overline{6}$	38)	$0.9\overline{3}$	39)	$0.9\overline{6}$	40)	$1.4\overline{6}$

Lesson 3 Converting fractions to terminating decimals. Review 1 A

1)	0.2	2)	0.4	3)	0.6	4)	0.8
5)	1.2	6)	1.4	7)	1.6	8)	1.8
9)	0.1	10)	0.3	11)	0.7	12)	0.9
13)	0.05	14)	0.15	15)	0.35	16)	0.45
17)	0.55	18)	0.65	19)	0.85	20)	0.95
21)	0.52	22)	0.68	23)	0.76	24)	0.92
25)	0.22	26)	0.46	27)	0.74	28)	0.98
29)	0.11	30)	0.23	31)	0.37	32)	0.49
33)	1.44	34)	2.64	35)	3.72	36)	4.76
37)	5.84	38)	6.88	39)	7.92	40)	8.96

Grade 7 Answers Chapter 5

Lesson 3 Converting decimals to fractions. Review 1 B

1)	$\frac{1}{5}$	2)	$\frac{2}{5}$	3)	$\frac{3}{5}$	4)	$\frac{4}{5}$	5)	$1\frac{2}{5}$
6)	$\frac{1}{4}$	7)	$\frac{3}{4}$	8)	$1\frac{1}{4}$	9)	$1\frac{3}{4}$	10)	$2\frac{1}{4}$
11)	$\frac{1}{8}$	12)	$\frac{3}{8}$	13)	$\frac{5}{8}$	14)	$\frac{7}{8}$	15)	$1\frac{1}{8}$
16)	$\frac{1}{16}$	17)	$\frac{3}{16}$	18)	$\frac{5}{16}$	19)	$\frac{7}{16}$	20)	$\frac{9}{16}$
21)	$\frac{1}{9}$	22)	$\frac{2}{9}$	23)	$\frac{1}{3}$	24)	$\frac{4}{9}$	25)	$\frac{5}{9}$
26)	$\frac{2}{3}$	27)	$\frac{7}{9}$	28)	$\frac{8}{9}$	29)	$\frac{2}{11}$	30)	$\frac{3}{11}$
31)	$\frac{4}{11}$	32)	$\frac{5}{11}$	33)	$\frac{6}{11}$	34)	$\frac{7}{11}$	35)	$\frac{8}{11}$
36)	$\frac{9}{11}$	37)	$\frac{4}{15}$	38)	$\frac{8}{15}$	39)	$\frac{11}{15}$	40)	$\frac{14}{15}$

Lesson 4 Operations with decimals and fractions. Review 1 A

1)	1.55	2)	1.45	3)	2.15	4)	1.975	5)	2.425
6)	1.375	7)	2.625	8)	0.625	9)	0.125	10)	0.15
11)	0.05	12)	0.15	13)	0.3	14)	0.8	15)	0.5
16)	3.65	17)	0.3	18)	0.15	19)	0.6	20)	0.3

Review 1 B

1)	1.8	2)	0.6	3)	2 2	4)	4.2	5)	0.9
6)	0.2	7)	0.8	8)	3.2	9)	0.8	10)	0.8
11)	0.8	12)	0.5	13)	1	14)	0.2	15)	0.6
16)	1.4	17)	0.6	18)	1.2	19)	1.5	20)	1.2

Grade 7 Answers Chapter 5

Lesson 5 Order of operations with decimals and fractions 1. R 1A

1)	3.5	2)	1.75	3)	2	4)	3	5)	0.2
6)	6	7)	0.5	8)	5.4	9)	9	10)	0.6
11)	7	12)	0.6	13)	2	14)	8	15)	0.8

Review 1 B

16)	0.5	17)	2	18)	2	19)	3	20)	0.2
21)	0.2	22)	2	23)	2	24)	0.2	25)	0.2
26)	0.4	27)	0.7	28)	2.5	29)	1.8	30)	6

Lesson 6 Order of operations 2. Review 1 A

1)	5	2)	1	3)	22	4)	11	5)	17
6)	5	7)	5	8)	9	9)	1	10)	11
11)	6.5	12)	6.5	13)	1.5	14)	13	15)	2.5
16)	20	17)	23	18)	9.25	19)	2.75	20)	2.5

Lesson 6 Review 1 B

1)	9	2)	6	3)	5.5	4)	1.5	5)	3.5
6)	3	7)	5	8)	3	9)	2.5	10)	6
11)	16	12)	1	13)	2.5	14)	1	15)	1.75
16)	2	17)	2	18)	2	19)	6	20)	6
21)	12	22)	0.5	23)	30	24)	4	25)	24
26)	3	27)	18	28)	15	29)	9	30)	16

Lesson 7 Operations and laws for numbers

Review 1 A

1) 0.5	2) 0.2	3) 1.5	4) 1.75	5) 1.5
6) 1	7) 0.75	8) 2	9) 1.5	10) 3

Review 1 B

1)	8	2)	1	3)	2.5	4)	1	5)	1.75
6)	1.5	7)	2	8)	3	9)	23	10)	37

Lesson 8 Worksheet 3 A for Final Exam

1) 0.25	2) 0.5	3) 0.1	4) 0.25	5) 0.5
6) 0.4	7) 0.8	8) 0.6	9) 0.2	10) 0.1

Worksheet 3 B for Final Exam

1) 3	2) 5	3) 2	4) 10	5) 6
6) 14	7) 26	8) 18	9) 12	10) 8

Grade 7 Answers Chapter 5

Lesson 9 Review for Test 5

Review 5 A

1) $\frac{1}{8}$ 2) 0.12 3) 1.375 4) 0.8 5) 0.6

6) 0.2 7) 1.2 8) 12 9) 3 10) 2

Review 5 B

1) $\frac{3}{8}$ 2) 0.28 3) 2.625 4) 0.5 5) 2

6) 0.6 7) 1.5 8) 0.25 9) 0.5 10) 3

Lesson 10 Test 5

Test 5 A

1) $\frac{5}{8}$ 2) 0.36 3) 1.75 4) 0.5 5) 0.9

6) 0.48 7) 2 8) 4 9) 3 10) 2

Test 5 B

1) $\frac{7}{8}$ 2) 0.44 3) 2.75 4) 0.5 5) 3

6) 0.76 7) 2.5 8) 6 9) 0.125 10) 3

Lesson 11 Review 7 for Final Exam (Basic Problems)

Review 7 A 1) 6 2) 1 3) 1.25 4) 16 5) 9 6) 0.125 7) 23
8) 3.5 9) 2 10) 26

Review 7 B 1) 70 2) 0.8 3) 2.5 4) 1 5) 15 6) 0.375 7) 9.25
8) 3 9) 3 10) 18

Lesson 12 Final Exam 7 (Basic Problems)

Test 7 A 1) 0 2) 3 3) 0.4 4) 2.5 5) 12 6) 0.625 7) 1.5
8) 5 9) 23 10) 12

Test 7 B 1) 50 2) 0.6 3) 2.5 4) 1 5) 6 6) 0.875 7) 13
8) 3 9) 37 10) 8

Grade 7 Chapter 1 Lesson 5

Table 1

Multiplication and Division

A + 0 = A 0 + A = A A – 0 = A A – A = 0

A × 1 = A 1 × A = A A ÷ 1 = A A ÷ A = 1

A × 0 = 0 0 × A = 0 0 ÷ A = 0 A ÷ 0 = Ø

A + B = B + A A × B = B × A Ø - no solutions

Number 5 from the left column multiply by number 4 from first row to get answer 20.

5 × 4 = 20 20 ÷ 5 = 4 20 ÷ 4 = 5

Memorize Table

×	1	2	3	4	5	6	7	8	9	10
1	1	2	3	4	5	6	7	8	9	10
2	2	4	6	8	10	12	14	16	18	20
3	3	6	9	12	15	18	21	24	27	30
4	4	8	12	16	20	24	28	32	36	40
5	5	10	15	20	25	30	35	40	45	50
6	6	12	18	24	30	36	42	48	54	60
7	7	14	21	28	35	42	49	56	63	70
8	8	16	24	32	40	48	56	64	72	80
9	9	18	27	36	45	54	63	72	81	90
10	10	20	30	40	50	60	70	80	90	100

Grade 7 Chapter 2 Lesson 3

Table 2
Perfect Square Numbers

Take number 20 from the left column add number 5 from the first row to get number 25
Answer is in intersection row of 20 and column of 5
$$25^2 = 25 \cdot 25 = 625$$

Memorize Table of Perfect Square for Indicated Numbers

a^2	0	1	2	3	4	5	6	7	8	9
10	100	121	144	169	196	225	256	289	324	361
20	400	441	484	529	576	625	676	729	784	841
30	900	961	1024	1089	1156	1225	1296	1369	1444	1521
40	1600	1681	1764	1849	1936	2025	2116	2209	2304	2401
50	2500	2601	2704	2809	2916	3025	3136	3249	3364	3481
60	3600	3721	3844	3969	4096	4225	4356	4489	4624	4761
70	4900	5041	5184	5329	5476	5625	5776	5929	6084	6241
80	6400	6561	6724	6889	7056	7225	7396	7569	7744	7921
90	8100	8281	8464	8649	8836	9025	9216	9409	9604	9801

Grade 7 Chapter 3 Lesson 5
Table 3 Decimals

Properties of Decimal

$4 = 4. = 4.0 = 4.00$　　$5.00 = 5$

Rounding of Decimals

$2.83 \approx 2.8$　　$2.38 \approx 2.4$

Comparison of Decimals

$1.48 < 3.6$　　$0.5 > 0.05$

Addition and Subtraction of Decimals

$$\begin{array}{r} 1.25 \\ +\,0.80 \\ \hline 2.05 \end{array} \qquad \begin{array}{r} 1.25 \\ -\,0.80 \\ \hline 0.45 \end{array}$$

Multiplication of Decimals

$0.24 \times 10 = 2.4$　　$0.24 \times 100 = 24$　　$0.24 \times 1000 = 240$

$2.5 \times 4 = 10$　　$1.25 \times 8 = 10$　　$0.5 \times 0.2 = 0.1$

Powers of Decimals

$3.5^2 = 3.5 \times 3.5 = 12.25$　　$0.5^3 = 0.5 \times 0.5 \times 0.5 = 0.25 \times 0.5 = 0.125$

Division of Decimals

$5 \div 10 = 0.5$　　$5 \div 100 = 0.05$　　$5 \div 1000 = 0.005$

$2 \div 5 = 0.4$　　$0.2 \div 5 = 0.04$　　$0.02 \div 5 = 0.004$

$2 \div 0.5 = 20 \div 5 = 4$　　$0.2 \div 0.5 = 2 \div 5 = 0.4$　　$0.02 \div 0.5 = 0.2 \div 5 = 0.04$

Order of Operations with the Same Level

$0.8 + 0.6 + 0.2 = 0.8 + 0.2 + 0.6 = 1.6$　　$2.5 \times 6 \times 4 = 2.5 \times 4 \times 6 = 60$

$0.8 - 0.2 + 0.4 = 0.6 + 0.4 = 1$　　$1.6 \div 4 \times 2 = 0.4 \times 2 = 0.8$

Order of Operations of Two Levels

$0.2 \times 4 - 0.2 = 0.8 - 0.2 = 0.6$　　$1.8 \div 9 + 0.4 = 0.2 + 0.4 = 0.6$

$0.6 + 0.4 \times 2 = 0.6 + 0.8 = 1.4$　　$0.8 - 1.5 \div 3 = 0.8 - 0.5 = 0.3$

Order of Operations in Expression with Parentheses

$(0.5 + 0.4) \times 2 = 0.9 \times 2 = 1.8$　　$3 \times (0.9 - 0.7) = 3 \times 0.2 = 0.6$

$(1.8 - 0.4) \div 7 = 1.4 \div 7 = 0.2$　　$2 \div (0.4 - 0.2) = 2 \div 0.2 = 10$

Groping Symbols:1) Parentheses ( )　2) Bracket [ ]　3) Brace { }

Order of Five Operations

Grouping Symbols (1)　[2]　{3}　　Exponents (Power) a^n

Multiplication or Division from left to right　　$\times$ or $\div$　$L \to R$

Addition or Subtraction from left to right　　$+$ or $-$　$L \to R$

Grade 7 Chapter 4 Lesson 2

Table 4

Greatest Common Factor (in short GCF)

The Greatest Common Factor is the greatest number by which all given numbers can be divided.

1) Find the GCF of numbers 12 and 18 (÷ 1, 2, 3, 4, ...)

Divisors of number 12 are 1, 2, 3, 4, 6, 12

Divisors of number 18 are 1, 2, 3, 6, 9, 18

Common divisors are 1, 2, 3, 6 Greatest common divisor is 6

GCD (12, 18) = 6 12 ÷ 6 = 2 18 ÷ 6 = 3

2)If greatest of the numbers can divided evenly by the smallest number, then the greatest common factor of numbers is equal to the smallest of given numbers: Find the GCF of numbers 12 and 4

Factors of 12 are 1, 2, 3, 4, 6, 12 Factors of 4 are 1, 2, 4

Common factors are 1, 2, 4 Greatest common factor is 4

GCF (12, 4) = 4 12 ÷ 4 = 3

3) The greatest common factor of prime numbers equal one:

GCF (5,7) = 1

Least Common Multiples (in short LCM)

The least number is the smallest number that can be evenly divided by each of the given numbers

1) Find the LCM of numbers 12 and 18 (×1, 2, 3, 4, ...)

Multiples of 12 are 12, 24, 36, 48, 60, 72...

Multiples of 18 are 18, 36, 54, 72...

Common multiples are 36, 72... Least common multiples is 36

LCM (12, 18) = 36 36 ÷ 12 = 3 36 ÷ 18 = 2

3)If the greatest of given numbers evenly divided by the smallest number, the LCM of the numbers is equal to the greatest of the given numbers: Find the LCM of numbers 4, 12

Multiples of 4 are 4, 8, 12, 16, 20, 24... Multiples of 12 are 12, 24...

Common multiples are 12, 24... LCM (4, 12) = 12

3)The least common multiple of prime numbers is equal to the product of the numbers. Find the LCM of numbers 2 and 5

Multiples of 2 are 2, 4, 6, 8, 10... Multiples of 5 are 5, 10...

Common multiples are 10, 20 ... LCM (2, 5) = 2 × 5 = 10

Grade 7 Chapter 4 Lesson 6

Table 5 Fractions

Properties of Fraction

$\frac{a}{b}=\frac{a\cdot n}{b\cdot n}$ $\qquad$ $\frac{a\div n}{b\div n}=\frac{a}{b}$

$\frac{2}{3}=\frac{2\times 6}{3\times 6}=\frac{12}{18}$ $\qquad$ $\frac{60}{75}=\frac{60\div 3}{75\div 3}=\frac{20}{25}=\frac{20\div 5}{25\div 5}=\frac{4}{5}$

Comparison of Fractions

$\frac{6}{11}>\frac{7}{22}$ $\qquad$ $\frac{6}{11}=\frac{6\cdot 2}{11\cdot 2}=\frac{12}{22}$ $\qquad$ $\frac{12}{22}>\frac{7}{22}$

Addition of Fractions

$\frac{a}{c}+\frac{b}{c}=\frac{a+b}{c}$ $\qquad$ $\frac{2}{5}+\frac{3}{8}+\frac{7}{20}=\frac{2\cdot 8}{5\cdot 8}+\frac{3\cdot 5}{8\cdot 5}+\frac{7\cdot 2}{20\cdot 2}=\frac{16}{40}+\frac{15}{40}+\frac{14}{40}=\frac{45}{40}=1\frac{1}{8}$

Subtraction of Fractions

$\frac{a}{c}-\frac{b}{c}=\frac{a-b}{c}$ $\qquad$ $\frac{3}{10}-\frac{1}{4}=\frac{6}{20}-\frac{5}{20}=\frac{1}{20}$ $\qquad$ $\frac{1}{12}-\frac{1}{18}=\frac{3}{36}-\frac{2}{36}=\frac{3-2}{36}=\frac{1}{36}$

Multiplication of Fractions

$\frac{a}{b}\times\frac{c}{d}=\frac{a\times c}{b\times d}=\frac{ac}{bd}$ $\qquad$ $\frac{4}{6}\times\frac{9}{8}=\frac{4\times 9}{6\times 8}=\frac{1\times 3}{2\times 2}=\frac{3}{4}$

$\frac{a}{b}\cdot c=\frac{a}{b}\cdot\frac{c}{1}=\frac{a\cdot c}{b}$ $\qquad$ $\frac{3}{8}\times 4=\frac{3}{8}\times\frac{4}{1}=\frac{3\times 1}{2\times 1}=\frac{3}{2}=1\frac{1}{2}$

$a\cdot\frac{b}{c}=\frac{a}{1}\cdot\frac{b}{c}=\frac{ab}{c}$ $\qquad$ $16\times\frac{5}{8}=\frac{16}{1}\times\frac{5}{8}=\frac{2\times 5}{1}=10$

Division of Fractions

$\frac{a}{b}\div\frac{c}{d}=\frac{a}{b}\cdot\frac{d}{c}=\frac{ad}{dc}$ $\qquad$ $\frac{5}{6}\div\frac{10}{3}=\frac{5}{6}\times\frac{3}{10}=\frac{1\times 1}{2\times 2}=\frac{1}{4}$

$\frac{a}{b}\div c=\frac{a}{b}\cdot\frac{1}{c}=\frac{a}{bc}$ $\qquad$ $\frac{3}{4}\div 6=\frac{3}{4}\div\frac{6}{1}=\frac{3}{4}\times\frac{1}{6}=\frac{1}{8}$

$a\div\frac{b}{c}=\frac{a}{1}\cdot\frac{c}{b}=\frac{ac}{b}$ $\qquad$ $6\div\frac{3}{4}=\frac{6}{1}\times\frac{4}{3}=\frac{8}{1}=8$

Power of Fractions

$\left(\frac{a}{b}\right)^n=\frac{a^n}{b^n}$ $\qquad$ $\left(\frac{1}{3}\right)^2=\frac{1^2}{3^2}=\frac{1}{9}$

Grade 7 Chapter 4 Lesson 6

Table 6 Mixed Numbers

Mixed Number $2+\frac{1}{2}=2\frac{1}{2}$ $\quad \frac{1}{4}+1=1\frac{1}{4}$ $\quad 1\frac{3}{4}=\frac{4\cdot 1+3}{4}=\frac{7}{4}$

Comparison of Mixed Numbers $1\frac{1}{3}<1\frac{5}{6}$ $\quad 1\frac{2}{6}<1\frac{5}{6}$

Addition of Mixed Numbers

$1\frac{1}{8}+2\frac{3}{8}=3\frac{4}{8}=3\frac{1}{2}$ $\quad 1\frac{1}{3}+2\frac{3}{9}=1\frac{1}{3}+2\frac{1}{3}=3\frac{2}{3}$

$1\frac{2}{6}+2\frac{1}{9}=1\frac{1}{3}+2\frac{1}{9}=1\frac{3}{9}+2\frac{1}{9}=3\frac{4}{9}$ $\quad 1\frac{1}{2}+1\frac{1}{3}+2\frac{1}{6}=1\frac{3}{6}+1\frac{2}{6}+2\frac{1}{6}=4\frac{6}{6}=5$

Subtraction of Mixed Numbers

$6\frac{7}{8}-5\frac{3}{8}=1\frac{4}{8}=1\frac{1}{2}$ $\quad 4\frac{3}{4}-2\frac{1}{4}=2\frac{2}{4}=2\frac{1}{2}$ $\quad 3\frac{1}{2}-2=1\frac{1}{2}$

$1-\frac{2}{3}=\frac{3}{3}-\frac{2}{3}=\frac{1}{3}$ $\quad 4-\frac{5}{8}=3\frac{8}{8}-\frac{5}{8}=3\frac{3}{8}$ $\quad 8-3\frac{5}{6}=7\frac{6}{6}-3\frac{5}{6}=4\frac{1}{6}$

$6\frac{2}{7}-2\frac{5}{7}=5\frac{9}{7}-2\frac{5}{7}=3\frac{4}{7}$ $\quad 5\frac{1}{3}-2\frac{5}{6}=5\frac{2}{6}-2\frac{5}{6}=4\frac{8}{6}-2\frac{5}{6}=2\frac{3}{6}=2\frac{1}{2}$

Multiplication of Mixed Numbers

$3\frac{1}{3}\cdot 2\frac{2}{5}=\frac{10}{3}\cdot\frac{12}{5}=\frac{2\cdot 4}{1\cdot 1}=8$ $\quad \frac{5}{8}\cdot 4\frac{4}{5}\cdot\frac{1}{6}=\frac{5}{8}\cdot\frac{24}{5}\cdot\frac{1}{6}=\frac{3}{6}=\frac{1}{2}$

Division of Mixed Numbers

$1\frac{1}{6}\div 4\frac{2}{3}=\frac{7}{6}\div\frac{14}{3}=\frac{7}{6}\times\frac{3}{14}=\frac{1}{4}$ $\quad 3\frac{1}{5}\div 2\frac{2}{5}=\frac{16}{5}\times\frac{5}{12}=\frac{4}{3}=1\frac{1}{3}$

$8\div 4\frac{4}{5}=\frac{8}{1}\div\frac{24}{5}=\frac{8}{1}\times\frac{5}{24}=\frac{5}{3}=1\frac{2}{3}$ $\quad 4\frac{1}{3}\div 13=\frac{13}{3}\div\frac{13}{1}=\frac{13}{3}\times\frac{1}{13}=\frac{1}{3}$

Power of Mixed Numbers

$\left(\frac{3}{4}\right)^3=\frac{3^3}{4^3}=\frac{27}{64}$ $\quad \left(1\frac{1}{2}\right)^2=\left(\frac{3}{2}\right)^2=\frac{3^2}{2^2}=\frac{9}{4}=2\frac{1}{4}$

Students!! Attention!!

Adults, help please students 1 – 4 grades to read and understand advice.

When I was in 5th grade I did not like math. I am not always understood math. I was 11 years girl. Next 6th grade I got a very good teacher. She explains very clearly and I understood her very well. I was worked very hard and I was not afraid to ask questions. Final Grade in Algebra in 6th grade I got an A. I said in class that I want to be a teacher of math. But some students said that I am not enough smart to be a math teacher. I said I will be work hard with math. I did it, I got a Master's degree in math and education, I love my students, I was teaching many years, in High School, part-time college math instructor, full-time university math lecturer.

Some people do a simple job, eat, sleep, go to work. People with a bad brain have not so a good job, not so good salary. If you look at TV how someone's dance, you will be not changing your brain. But if you take a dancing class this helps your body and brain ability. The same if you want to use the calculator, but the calculator not always with you. Much better if you learn how you need to do any operations with numbers without a calculator. If you read this math book you will know what you need to do. This teaches you how to improve your memory, attention, logic thinking, how to read your math books every day, your brain works better. You can work 50 minutes with math, 10 minutes break maybe dance. Do not afraid to do mistakes, fixed and move to the next step. Don't listen to anyone, choose what do you like more.

Students need to know if you work with the brain every day, your brain works better. You need a good brain in any job, in any class. When you finish High School, you need to get any profession, you need to go to college and get a good job and a better salary. You need to learn any subject, its work for the brain. Read the book, do the exercises yourself, try your best, and then check your answer at the end of the chapter. Be honest with yourself. You need to believe in yourself. The brain is like muscles. People with good brain lives longer. The professors at my university were over 80 years old, wise, and optimistic. It was a great pleasure to talk to them.

Made in United States
Orlando, FL
05 November 2026

90230214R00122